EXPERT SYSTEMS for Civil Engineers:
Education

Prepared by the Committee on Expert Systems
of the Technical Council on Computer Practices
of the American Society of Civil Engineers

Edited by Satish Mohan and Mary Lou Maher

Published by the
American Society of Civil Engineers
345 East 47th Street
New York, New York 10017-2398

ABSTRACT

While the technology of expert systems can provide engineers with another approach to the development of computable models for civil engineering problems, it is not clear what a course on expert systems for civil engineers should contain or how it should be taught. This book, which is comprised of three parts, begins to address these issues. Part one presents case studies of existing courses that provide several different perspectives on the content and presentation of a course. Part two is a collection of reference materials with commentaries on their appropriateness. The reference materials include books, articles, software and hardware. Finally, part three contains listings of various instructional aids such as assignments, projects, and examination questions. Although the book focuses on courses taught at universities, it does address some of the issues in developing continuing education courses for civil engineers.

Library of Congress Cataloging-in-Publication Data

Expert systems for civil engineers: education/prepared by the Committee on Expert Systems of the Technical Council on Computer Practices of the American Society of Civil Engineers; edited by Satish Mohan and Mary Lou Maher.
    p. cm.
    Includes bibliographical references.
    ISBN 0-87262-741-1
    1. Civil engineering—Data processing—Study and teaching (Higher)—United States. 2. Expert systems (Computer science)—Study and teaching (Higher)—United States. 3. Civil engineers—Training of—United States. I. Mohan, Satish. II. Maher, Mary Lou. III. American Society of Civil Engineers. Technical Council on Computer Practices. Committee on Expert Systems.
T73.E96   1989
624'.0285—dc20   89-18399
                                                                      CIP

The Society is not responsible for any statements made or opinions expressed in its publications.

Authorization to photocopy material for internal or personal use under circumstances not falling within the fair use provisions of the Copyright Act is granted by ASCE to libraries and other users registered with the Copyright Clearance Center (CCC) Transactional Reporting Service, provided that the base fee of $1.00 per article plus $.15 per page is paid directly to CCC, 27 Congress Street, Salem, MA 01970. The identification for ASCE Books is 0-87262/88. $1 + .15. Requests for special permission or bulk copying should be addressed to Reprints/Permissions Department.

Copyright © 1989 by the American Society of Civil Engineers,
All Rights Reserved,
Library of Congress Catalog Card No: 89-18399
ISBN 0-87262-741-1
Manufactured in the United States of America.

FOREWORD

In the last five years, Expert Systems have become a subject of considerable coverage in the journals and professional meetings of the American Society of Civil Engineers (ASCE). Much of the credit for this exposure is due to the active program of promulgation and discussion by the Expert Systems Committee of the Technical Council on Computer Practices (TCCP).

Why is the subject of expert systems so popular? Since the inception of ASCE's computer-related activities 32 years ago, when the Structural Division's Committee on Electronic Computation, was formed no single topic - not even CAD or PC's - has elicited the same level of interest. The answer is, undoubtedly, that we are engineers and not scientists. As engineers, we realize that all knowledge does not come from textbooks, and that engineering solutions involve more than the application of formulas. Expert systems promise to provide a methodology whereby the *heuristics* of experiential knowledge can be captured and applied, thereby elevating the assistance provided by computers from *calculating* to *reasoning*.

This monograph is predicated on the belief by the Expert Systems Committee that education holds the key to the wide-scale incorporation of expert systems into all aspects of civil engineering practice. The Committee has done a masterful job of collecting information about existing expert systems courses in civil engineering and compiling it in a form useful for instructors planning to offer such courses. Instructors presently teaching such courses, as well as individuals and organizations developing or planning to develop expert systems programs will also find much useful information in this monograph.

The majority of courses described in this monograph are offered at the graduate, rather than the undergraduate level. There are a number of reasons for this fact: new subjects typically start as graduate level courses before they are incorporated into the undergraduate curriculum; the undergraduate program is already bulging with required courses, etc. However, there is a deeper reason: these courses are aimed at civil engineers, not computer science majors, and all of the courses expect students to build their own, albeit prototype, expert systems as part of the course requirements. This requires some level of expertise in *some* aspect of civil engineering, and it is this level of expertise that undergraduate students do not yet possess. There is no doubt that the methodological aspects of expert systems can be "down-sized" for undergraduate coverage; it is an open question whether a suitable experience base can be found for these students.

It is my hope that this monograph will encourage a number of civil engineering faculty members to offer a course on expert systems, and that a course on expert systems will become an integral part of most civil engineering curricula. I have found it most rewarding to teach such a course; the breadth of topics chosen by students for expert system implementation continually amazes me. I trust that others will find the subject equally exciting.

Steven J. Fenves, Hon. M. ASCE
Pittsburgh, Pennsylvania
July, 1989

Preface

Expert system technology has provided another approach to developing computable models for civil engineering problems. As such, the concepts and applications of this technology have become another set of tools available to civil engineers. Understanding and learning to use these tools is a problem that faces civil engineers as they try to keep up with current developments in computing and engineering. However, it is not clear what a course on expert systems for civil engineers should contain and how it should be taught. These are issues faced by both academics maintaining a curriculum for undergraduate and graduate civil engineering programs and practicing engineers that need to keep up with new technology. This book begins to address some of the issues in teaching expert systems to civil engineers by presenting case studies of existing courses and course material for use in preparing a course. Although the book focusses on courses taught at universities, it does address some of the issues in developing short courses directed towards the continuing education of civil engineers.

The book is divided into three parts:
1. Part I: Expert Systems Courses,
2. Part II: Expert Systems Teaching Aids and Materials, and
3. Part III: Sample Homework Assignments, Class Projects and Examinations.

The first part is a selection of case studies of existing expert systems courses taught by civil engineering professors, usually within a civil engineering graduate program. The case studies include several different approaches and objectives for teaching expert systems to civil engineers. The second part includes reference material for use in an expert systems course, guidance on the appropriateness of the different reference materials, and the use of the various hardware and software configurations. The third part is a listing of the more civil engineering oriented assignments and examination questions selected among those submitted by the instructors.

The material presented in this book is predicated on the assumption that civil engineers should be teaching courses on expert systems to civil engineering practitioners and students. Although this is still a controversial subject, it is generally agreed that civil engineers should know enough about this technology to take advantage of it in their work. In this book we hope to facilitate the development of courses on expert systems with the broader goal of making the technology more available to civil engineers. In developing a new course based on a technology from computer science, one of the more difficult challenges is identifying appropriate pedagogical aids and assignments. It is our intent that the reader of this book will benefit from the experience of others in teaching the concepts and applications of expert systems to civil engineers.

Satish Mohan and Mary Lou Maher, Editors

Acknowledgments

This book is an effort of the ASCE Expert Systems Committee to disseminate information on expert systems to the civil engineering community. The book is second in a series, where each book deals with a specific issue related to expert systems. The first book, **Expert Systems For Civil Engineers: Technology and Application** was published by ASCE in 1987. These books represent an ongoing effort of the committee to prepare and carefully monitor a set of publications directly related to civil engineering and expert systems. Each chapter contributed to these books is reviewed by the committee for scope and content. The editors appreciate the efforts of the entire committee in guiding the development of this monograph.

Members of the Expert Systems Committee

Robert Allen
Tomasz Arciszewski
David B. Ashley
William Bowlby
Louis F. Cohn
Jesus M. De la Garza
Michael Demetsky
Clive L. Dym
Gavin A. Finn
Bruno M. Franck
Jon Fricker
James Garrett
John S. Gero
Geoffrey D. Gosling
Roswell A. Harris
Craig Howard
William Ibbs
R. Raymond Issa
Simon Kim
Kincho H. Law
Ray Levitt
Andrew B. Levy

Phillip J. Ludvigsen
Mary Lou Maher
Satish Mohan
Peter Mullarkey
Glenn Orenstein
Leonard Ortolano
Richard N. Palmer
William Rasdorf
Dorothy A. Reed
Stephen G. Ritchie
W.M. Kim Roddis
Paul N. Roschke
Timothy J. Ross
Lewis A. Rossman
Thomas V. Schields
Thomas Siller
Miroslaw J. Skibniewski
D. Sriram
Kenneth M. Strzepek
Charles H. Trautmann
Terence A. Weigel

Biographies Of Authors

Tomasz Arciszewski is an Associate Professor of Civil Engineering at Wayne State University in Detroit. His major research area is design methodology and the applications of artificial intelligence in engineering. He has authored or co-authored more than 40 publications in the areas of structural engineering, design methodology, and expert systems. He founded the Intelligent Computers Laboratory at WSU in 1987. The mission of the laboratory is to do research on expert systems and knowledge acquisition in civil engineering, including inductive learning and its engineering methodology.

Demetre P. Argialas received the B.S., M.S., and Ph.D. degrees from the National Technical University of Athens, the University of Tennessee, and the Ohio State University, respectively. He is currently an Assistant Professor of Civil Engineering, and member of the Remote Sensing and Image Processing Laboratory at Louisiana State University. His research and teaching interests are terrain analysis, remote sensing, digital image analysis, knowledge based expert systems, and digital mapping. He is a member of ASCE, ASPRS, IEEE, AAAI, ACSM, and TRB.

James H. Garrett, Jr. is currently an assistant professor of Civil Engineering at the University of Illinois at Urbana-Champaign. His research interests include: object-oriented representations of design standards and building information, knowledge-based usage of design standards for both conformance checking and design, and neural network approaches to structural and environmental classification problems. Professor Garrett received his Ph.D. in 1986 from Carnegie Mellon University in Pittsburgh. After receiving his Ph.D., he joined Schlumberger Well Services-Houston Downhole Sensors in Houston, Texas, where he worked for a year on knowledge-based systems to perform electronic power transformer design and material selection. Dr. Garrett is a recent recipient of an NSF Presidential Young Investigator Award.

Steven J. Fenves is the SUN Company University Professor of Civil Engineering at Carnegie Mellon University. He received his B.S., M.S., and Ph.D. degrees in Civil Engineering from the University of Illinois, where he served on the faculty until 1971. Since he joined Carnegie Mellon University in 1972, he has served as Head of the Department of Civil Engineering and Director of the Design Research Center. His work in computer-aided design began in 1962 with the development of the STRESS systems. He has been active in several aspects of computer-aided engineering, including network-based algorithms, problem-oriented languages, representation and processing of specifications, engineering databases, integrated design systems, and engineering design applications of knowledge-based expert systems and artificial intelligence.

Philip Ludvigsen directs the design, development, and delivery of environmental decision support systems at Environmental Resources Management, Inc. His work involves the integration of expert systems, data bases and hypermedia into deliverable systems designed to meet the needs of environmental decision makers. Dr. Ludvigsen has managed or advised on the development of decision support systems for EPA, Environment Canada, and the American Petroleum Institute. Other interests include interface design, computer-based training, and computer assisted risk assessment.

Mary Lou Maher is an Associate Professor in the Civil Engineering Department at Carnegie Mellon University. She received the B.S.C.E. from Columbia University in 1979, and the M.S. and Ph.D. from Carnegie Mellon in 1981 and 1984, respectively. Her research includes the use of knowledge based expert system techniques in computer aided design applications, specifically in structural design. Dr. Maher was awarded Carnegie Mellon's Ladd research award and the National Science Foundation's Presidential Young Investigator Award in 1987. She is Chairman of the Expert Systems Committee of the American Society of Civil Engineers.

Satish Mohan is an Associate Professor of Civil Engineering and Coordinator of Grduate Construction Management Program at the State University of New York at Buffalo. He received the B. Tech. (Hons) degree in Civil Engineering from the Indian Institute of Technology, Khargpur, India, the M.S. degree from Kansas State University in 1975 and Ph.D. from Purdue University in 1978. He has worked on the planning design and construction of roads, bridges, buildings, and an airport in India and Tanzania. His research interests include: expert systems, computer applications in construction, decision processes, statistical quality control, and traffic safety.

John M. Niedzwecki is currently an Associate professor of Civil Engineering at Texas A&M University in College Station, Texas. He is a member of their Ocean Engineering Program. Dr. Niedzwecki is a registered professional engineer and is active in ASCE. His teaching and externally funded research activities have been primarily focused upon offshore engineering mechanics. Currently, he is chairman of the department's committee on Computer Systems and Applications and, is directing the development of an advanced computational laboratory for the Department of Civil Engineering.

Richard N. Palmer is an Associate Professor of Civil Engineering at the University of Washington, Seattle, Washington. He received his B.S. at Lamar University, M.S. from Stanford University, and his Ph.D. at John Hopkins University. His research interests include water resources management, risk analysis, natural language processing, and the application of expert systems to a wide range of Civil Engineering problems. He is an Associate Editor of the Journal of Water Resources Planning and Management of ASCE and on the Editorial Board of Civil Engineering Systems.

Duvvuru Sriram is an Assistant Professor of Civil Engineering and co-technical director of the Intelligent Engineering Systems Laboratory at M.I.T. Currently, he is working in the areas of knowledge-based systems, geometrical reasoning and object-oriented databases. He has nearly 50 publications to his credit, including five books. He was a founding co-editor of the International Journal for AI in Engineering. In 1989, he was awarded the Presidential Young Investigator's award from the National Science Foundation. Sriram received a B. Tech (1980) degree from I.I.T., Madras, India, and MS (1981) and Ph.D. (1986) degrees from Carnegie Mellon University.

Jeff R. Wright joined the faculty of the School of Civil Engineering at Purdue University in the Fall of 1982 after completion of his Doctorate from the John Hopkins University. Previous degrees include Bachelor's degree in Social Psychology and Engineering, and a Master of Science in Civil Engineering from the University of Washington. Professor Wright has developed an active research program focusing on the support of the engineering function by modern computer technology. His interests in this area range from simulation and optimization modeling of large-scale engineering systems to automatic data collection and analysis methods. Most of his research to date has focused on water and energy systems.

List of Software Programs

The following software programs are referred to within the text:

Name:	*Developed or Marketed By:*
The Deciding Factor	Channelmark Corp
TAX	D. Argialas, Louisiana State University
Insight 2+	Level 5 Research, Indinapolis, Florida
KEE	IntelliCorp, Mountain View, CA
OPSS	Garnegie-Mellon University, Pittsburgh, PA
Turbo Prolog	Borland International, Scotts Valley, CA
VP Expert	Paperback Software
XLISP	D. Betz, Manchester, N.H.
Intelligence Compiler	Inteligence Ware, Inc., Los Anaeles, CA

Table of Contents

Foreword	by Steven J. Fenves	III
Preface		v
Acknowledgements		vii
Contributors		ix
List of Software Programs		xi

Part I: Expert Systems Courses

Chapter 1: **Expert Systems Education in Civil Engineering - State of the Art**
by Satish Mohan 1

Chapter 2: **Teaching Expert Systems Techniques at Wayne State University**
by Tomasz Arciszewski 15

Chapter 3: **Teaching Expert Systems Techniques at Louisiana State University**
by Demetre P. Argialas 23

Chapter 4: **A Course on Expert Systems in Civil Engineering at Carnegie Mellon University**
by Steven Fenves 30

Chapter 5: **Teaching a Graduate Course on Expert Systems to Civil Engineers**
by Satish Mohan 33

Chapter 6: **A Course on Expert Systems in Civil Engineering**
by John M. Niedzwecki 35

Chapter 7: **Teaching Expert Systems to Civil Engineers at the University of Washington: An Opportunity for Assessment**
Richard N. Palmer 45

Chapter 8: **Teaching Expert Systems Techniques at M.I.T.**
by D. Sriram 54

Chapter 9: **Teaching Expert Systems Techniques at Purdue**
by Jeff R. Wright 60

Chapter 10: **Teaching an Expert Systems Short Course, A Personal Perspective**
by Richard N. Palmer 70

Chapter 11: **The ASCE Short Course on Expert Systems**
by Mary Lou Maher 79

Part II: Expert Systems Teaching Aids and Materials

Chapter 12: **Texts and Reference Books on Expert Systems and Artificial Intelligence**
by James H. Garrett, Jr. 82

Chapter 13: **Knowledge-Based Expert Systems in Civil Engineering: An Annotated Bibliography**
by D. Sriram 95

Chapter 14:   **Hardware and Software for Expert**
                Systems Courses
                by Philip Ludvigsen and Mary Lou Maher 112

Part III: Sample Homework Assignements, Class Projects
                        and Examinations

Chapter 15:   **Selected Homework Assignments** 117
Chapter 16:   **Selected Project Assignments** 129
Chapter 17:   **Selelected Examination Questions** 133

Chapter 1

Expert Systems Education in Civil Engineering State-of-the-Art

by Satish Mohan, M. ASCE

1 INTRODUCTION

At this time, there are very few operational expert systems in civil engineering. One possible reason for this is the lack of formal education in expert systems to civil engineers. In most of the universities, expert systems techniques are taught by the computer science departments. These courses lay more emphasis on theoretical concepts rather than applications and often require one or two computer science courses as prerequisites; these conditions deter the civil engineering students from taking these courses. In the last few years, some of the civil engineering departments in the U.S.A. and abroad have started to teach expert systems techniques within the department which has generated an interest in expert systems techniques within the civil engineering community. A good number of expert system prototypes, in several civil engineering domains, have been developed in these courses; some of which can be advanced to operational status with user-industry support. The positive results of these courses have led many to believe that more of the civil engineering departments should offer courses in expert systems techniques so that full benefits of this technology may be realized by the civil engineering profession.

With a view to popularize, facilitate and enrich expert systems education in civil engineering, the Expert Systems Committee of the Technical Council on Computer Practices (TCCP) of the American Society of Civil Engineers (ASCE) decided to compile, evaluate and publish state-of-the-art teaching in expert systems. A survey of several universities in the U.S.A. and abroad was conducted for this purpose; this chapter summarizes the results of the survey.

One-hundred (100) of the civil engineering departments responded to the survey. Out of these, seventy-five (75) responses were from the U.S.A. and twenty-five (25) from abroad. These responses show that fifteen (15) of the universities are currently offering a course on expert systems within their civil engineering departments, and fourteen (14) others are planning to start offering a course on expert systems in the near future. The focus of expert system courses has been the development of skills to build expert system applications in civil engineering rather than the theoretical concepts; every course has included the development of a prototype system, which has been weighted heavily in grading.

There are many books available on expert systems but none of them matches the philosophy or course contents followed by the instructors. Most of the instructors use one or two of the available books and require supplemental reading from technical journals. Most recommended books and technical papers are listed in subsequent sections of this chapter.

Choice of computing environment is one of the issues raised by the instructors in their responses. The wide range of expert system software and hardware currently available in the market compounds the problem. Most of the courses, so far, have utilized PC-based shells. However, respondents suggest that an expert system software that supports hybrid programming environments would be better suited. This chapter lists ten most used expert system software.

In all of the civil engineering departments where courses in expert system techniques have been offered, students have shown a keen interest in the technology and willingly devoted more effort than other courses. Many instructors have reported that the results have been very satisfying -- both to them as well as to the students. During the process of building expert systems, students learn to understand and appreciate the value of professional experience and expertise in the civil engineering world.

Teaching an expert systems course has not been easy, according to the course instructors. Selection of books and reading material; selection, acquisition, and management of computing environment; learning new shells; and developing domain-specific problems for the class have been difficult. It is hoped that this monograph will help in resolving these and other problems.

2 SURVEY OF CURRENT TEACHING PRACTICES

To survey the expert systems teaching practices in civil engineering, a survey form as in the Appendix, was mailed to the following:

(i) Chairmen of civil engineering departments of universities in U.S.A., Canada, U.K., Australia, Europe, Africa, India, Southeast Asia and Japan.
(ii) Educators who were known to be involved in teaching expert systems.
(iii) Members of the ASCE TCCP Committee on Expert Systems.
(iv) Members of the Transportation Research Board (TRB) Task Force on Expert Systems.
(v) Members of the ASCE Technical Committee on Microcomputers in Construction.

A total of about 300 survey forms were mailed to the members of the above groups during the early months of 1988. One hundred (100) of the civil engineering departments responded to the survey. Out of these, seventy-five (75) responses were from U.S. and twenty-five (25) from abroad. The following statistics were taken from these responses.

(i) Fifteen (15) of the universities are currently offering a graduate and/or undergraduate course on expert systems.
(ii) Fourteen (14) of the universities are planning to start offering a course on expert systems in the near future.
(iii) Fifty-one (51) of the universities have no plans, but feel that offering a course in expert systems would be useful.
(iv) Nine (9) of the universities feel that an expert systems course is not necessary in a civil engineering curriculum.
(v) Six (6) of the universities are currently teaching expert systems techniques to civil engineering students in ways other than a formal course.
(vi) Five (5) of the responses received did not relate to the context of this monograph: the courses were not offered to civil engineering students, or the instructors moved to another university. These five responses have, therefore, not been included in any of the analyses presented in this chapter.

Fifteen (15) universities are currently offering expert systems courses in their civil engineering departments. Thirteen (13) of these universities are in U.S.A. and two (2) are abroad. These universities are listed in Table 1. A majority of the top 20 civil engineering departments in the U.S.A. are not offering an expert systems course. Except one university where this course started in 1985, most other courses started in 1986 or later.

Fourteen (14) universities are planning to start offering an expert systems course in civil engineering in the near future. One of these universities is in Australia, five in U.K., one in Switzerland, and seven in U.S.A. These universities are listed in Table 2. As indicated in the 'Remarks' column of this table, the teaching plans include several modes: five of them will teach expert systems as part of an undergraduate course, one of them will offer a graduate course, and one other as an AI applications course. Two of these 14 universities have recognized the potential benefits of expert systems in civil engineering applications, have already set up AI labs, and are doing graduate research at the present time. They have been waiting for expert systems technology to be readily accessible and to be further developed before they start offering regular undergraduate and graduate courses. These universities, and others in a similar dilemma, would greatly benefit in drawing upon the experience of those who have now been teaching expert systems techniques for several years and have seen their students apply the technology in solving civil engineering problems.

In response to question number (1) of the survey, fifty-one universities reported that they have no plans to offer an expert systems course in civil engineering but feel that such a course, if offered, would be useful. Thirty-seven (37) of these responses were from U.S. and fourteen (14) from abroad. Of those from abroad, three were from Canada; three from South Africa; two each from U.K., France, and Australia; and one each from India and the Federal Republic of Germany.

TABLE 1-1: Universities Currently Offering Expert Systems Courses in Civil Engineering

	University	Remarks
1.	University of California Irvine, CA 92717	· Graduate course. · Offered every winter quarter since 1986. · Prerequisite: none.
2.	Carnegie Mellon University Pittsburgh, PA 15213	· Graduate course, open to senior undergraduates. · Core course in the CAE graduate program in Civil Engineering. · Offered every spring semester since 1985. · Prerequisite: none, requires familiarity with CMU's computing environment.
3.	Georgia Institute of Technology Atlanta, GA 30332	· Graduate course. · Offered spring quarter since 1986. · Prerequisite: knowledge of microcomputers.
4.	Louisiana State University Baton Rouge, LA 70803	· Graduate course; future plans: an undergraduate course · First offered in fall 1987.
5.	Massachusetts Institute of Technology Cambridge, MA 02139	· Graduate course, open to all engineering disciplines. · Core course for "Intelligent Systems" majors. · Offered every fall since 1986.
6.	University of Massachusetts Amherst, MA 01003	· Graduate/undergraduate course.
7.	State University of New York (SUNY) at Buffalo Buffalo, NY 14260	· Graduate course. · Offered every spring since 1988. · Prerequisite: none.
8.	North Carolina State University Raleigh, NC 27695	· Graduate/undergraduate course. · Prerequisite: none.
9.	Purdue University W. Lafayette, IN 47907	· Primary focus: graduate students, open to senior undergraduates.
10.	Stanford University Stanford, CA 94305	· Graduate course. · Offered every spring quarter since 1986. · Prerequisite: none.
11.	Texas A & M University College Station, TX 77843	· Primary focus senior undergraduates, open to graduates, visitors and faculty. · First offered: spring 1986. · Prerequisite: none, assumes programming experience in FORTRAN.
12.	University of Washington Seattle, WA 98195	· Graduate & senior undergraduates. · First offered in winter 1986. · Prerequisite: one programming language.
13.	Wayne State University Detroit, MI 48202	· Graduate course, distributed in two courses. · First offered in winter 1986.
14.	Ecole Nationale Des Ponts et Chanssees 75007 Paris, France	· First offered in fall 1988. · Prerequisite: none.
15.	Chalmers University of Technology Goteborg, Sweden	· Prerequisite: One CS course.

TABLE 1-2: Universities Planning to Offer a Course on Expert Systems in the Department of Civil Engineering

University	Remarks
1. Monash University Clayton, Australia	
2. University of Bradford Bradford, BD7IDP, UK	To be taught as part of another course.
3. University College London WC1E6BT, UK	To be taught as part of Operations Research course.
4. Queen Mary College University of London London E14NS, UK	To be developed within an undergraduate course in Finite Element Method.
5. University of Salford Salford M54WT, UK	Currently teaching some expert system applications using ELSIE.™
6. Heriot-Watt University Riccarton, Edinburgh EH 144AS, UK	Have acquired AI hardware and software.
7. Swiss Federal Institute of Technology Lausanne, Switzerland	
8. University of Maryland College Park, MD 20742	
9. Florida International University Miami, FL 33199	
10. Washington University St. Louis, MO 63130	Graduate course starting in fall 1988.
11. Colorado State University Ft. Collins, CO 80523	Planning an AI applications course.
12. University of Wisconsin Madison, WI 53706	To be taught within a computing for engineers undergraduate course.
13. University of Colorado Boulder, CO 80309	
14. Vanderbilt University Nashville, TN 37235	To be taught within a 'Computer-Aided Civil Engineering' course.

A few of the respondents are concerned about the usefulness of expert systems technology unless efficient ways are found to systematize domain knowledge. Others have commented that they have, so far, not noticed any progressive use of expert systems technology in civil engineering. Most of these respondents are concerned about 'how to prepare potential instructors', 'what to teach', 'what should be the computing environment', and 'which books, references, etc. should be used'.

Nine (9) of the universities responded that an expert systems course is not necessary in a civil engineering curriculum. All of these nine responses were from the civil engineering departments of U.S. universities. Four of the nine universities indicated that their response relates to an already full undergraduate curriculum. Two of the nine universities have included expert systems techniques within other civil engineering courses and one of the nine sends their civil engineering students to electrical engineering and computer science departments for expert systems courses. The results of these nine responses are summarized as below:

(i) Two of the universities feel that an expert systems course is not necessary in a civil engineering curriculum both at the graduate level as well as at the undergraduate level,

(ii) Four of the universities feel that an expert systems course is not necessary at the undergraduate level, and

(iii) Three of the universities recognize the necessity of expert systems education but do not feel that a separate course should be offered within the civil engineering department.

Six (6) of the universities teach expert systems techniques using other approaches. Some of them teach within other civil engineering courses or as short courses open to students, faculty, and visitors. Some others send civil engineering students to another department to learn expert systems techniques. These six (6) universities are listed in Table 3. Four of these universities are in the U.S.A.

TABLE 1-3: Other Approaches to Teaching Expert Systems to Civil Engineers

	University	Teaching Mode
1.	University of Minnesota Minnesota, MN 55455	Expert systems taught within a geo-engineering undergraduate course in Systems Analysis, accessible to civil engineering undergraduates.
2.	University of Virginia Charlottesville, VA 22901	Civil engineering students take expert systems course offered by the Systems Engineering department.
3.	University of Illinois Urbana, IL 61801	Civil engineering students take an undergraduate course, 'Knowledge-Based Expert Systems for Engineers', offered by the mechanical engineering department.
4.	Florida Institute of Technology Melbourne, FL 32901	Civil engineering students take expert systems course offered by the industrial engineering department.
5.	Fachhochschule Rheinland-Pfalz Abteilung Mainz I, D-6500 Mainz, West Germany	Expert Systems concepts covered in a course 'CAD, Traffic Planning'.
6.	University of Edinburgh Edinburgh, EH9 3JL, UK	A short course on expert systems to senior undergraduates.

TABLE 1-4: Topics Covered in Expert Systems Courses

	Topic	No. of Lectures		Topic	No. of Lectures
1.	Overview of Artificial Intelligence (AI)	1	9.	Knowledge Acquisition - Elicitation methods - Consistency checks - Multiple experts	2-3
2.	Introduction to Knowledge Based Expert Systems (KBES) - Expert Systems versus Conventional Programs	1-2	10.	KBES Building, Testing, and Validation	2-3
3.	Major Expert System Applications	1-2	11.	Expert Systems Applications in Civil Engineering - Case studies	1-3
4.	Knowledge Representation: Rules, Frames, Logic, Semantic Nets	2-4	12.	KBES Integration - Algorithmic programs - Databases - Graphics packages	2-3
5.	Expert Systems Architecture - Production systems - Blackboard architectures	2-3	13.	Object Oriented Programming	1-2
6.	Inferencing Mechanisms - Control Strategies - Search Methods	2-4	14.	Advanced Problem Solving Strategies	1-2
7.	Management of Uncertainty & Fuzzy Knowledge	1-3	15.	AI Programming Languages: LISP, PROLOG, SMALLTALK^{TM}	3
			16.	Neural Networks	1-2
8.	Survey of Expert System Building Languages & Tools	1-2	17.	Future of AI and Expert Systems	1
			18.	Project Presentations	2-3

3 COURSE SYLLABUS

The focus of expert systems courses has been the development of skills to build expert system applications in civil engineering rather than the theoretical concepts. Table 4 gives a set of topics that have been included in most of the courses. The selection of topics and number of lectures devoted to each topic has varied with: (i) the background and current interests of the instructor, (ii) the length of the term, i.e., a 16-week semester or a 10-week quarter, and (iii) the expert system software/language being used in the course lab. For example, one university includes lectures and examples on 'inductive learning', another gives six out of 28 lectures on OPS5, and another devotes one-third of the course to the syntax of the expert system shell and its use for prototype development.

The sequencing of various topics has not been significantly different except that some of the instructors have presented expert system application examples quite early in the course. Some others have introduced AI languages in the first week.

4 BOOKS AND TECHNICAL JOURNALS

Learning expert systems concepts and techniques involves extensive reading from books and technical journals and building one or more prototype expert systems using commercially available software. Most of the universities include these readings on the exams. One university gives an oral exam on suggested readings which counts 30% towards their course grade.

There are many books on expert systems currently available but none of them matches the philosophy or the course contents followed by most of the instructors. At this time, some of the instructors have packaged their course notes for students' use, while others have prescribed one or two of the available books as texts and two to six of them as references. A list of ten most recommended text and reference books is given in Table 5.

TABLE 1-5: Ten Most Recommended Text and Reference Books

1.	Harmon, P., and King, D. *Expert Systems: Artificial Intelligence in Business*, Wiley, New York, 1985.	(12)*
2.	Hayes-Roth, F., Waterman, D.A., and Lenat, D.B., (editors), *Building Expert Systems*, Addison-Wesley, Reading, MA, 1983.	(7)*
3.	Waterman, D.A., *A Guide to Expert Systems*, Addison-Wesley, Reading, MA, 1986.	(7)*
4.	Charniak, E., and McDermott, D., *Introduction to Artificial Intelligence*, Addison-Wesley, Reading, MA, 1985.	(5)*
5.	Jackson, P., *Introduction to Expert Systems*, Addison-Wesley, Reading, MA, 1986.	(5)*
6.	Winston, P., *Artificial Intelligence - Second Edition*, Adison-Wesley, Reading, MA, 1985.	(5)*
7.	Brownston, L., Farrell R., Kant, E., and Martin, N., *Programming Expert Systems in OPS5*, Addison-Wesley, Reading, MA, 1985.	(4)*
8.	Kostem, C., and Maher, M.L., (editors), *Expert Systems in Civil Engineering*, American Society of Civil Engineers, New York, 1986.	(4)*
9.	Maher, M.L., (editor), *Expert Systems for Civil Engineers: Technology and Application*, American Society of Civil Engineers, New York, 1987.	(4)*
10.	Adeli, H., (editor), *Microcomputer Knowledge-Based Expert Systems in Civil Engineering*, American Society of Civil Engineers, New York, 1988.	(3)*

* Number of universities recommending the book

Almost all of the course syllabi refer to 15 to 30 technical papers from various journals for supplemental reading. Some of the instructors allocate 10-15 minutes of class time every week for discussions on that week's paper(s) and require written submissions of students' critique on the content and quality of the paper. The selection of technical papers for reading assignments varies according to the background and current emphasis of the instructor. Some of the universities make available to their students a collection of outstanding projects from previous years and their students have found them very useful. A list of five most recommended technical journals is given in Table 6 and a list of 20 most recommended technical papers is given in Table 7. Table 7 list does not include more recommended chapters from books listed in Table 5.

TABLE 1-6: Five Most Recommended Technical Journals

1.	*Journal of Computing in Civil Engineering*, American Society of Civil Engineers, 345 East 47th Street, New York, NY 10017-2398.	(2)*
2.	*The AI Magazine*, American Association of Artificial Intelligence, 445 Burgess Drive, Menlo Park, CA 94025.	(2)*
3.	*IEEE Expert*, Institution of Electrical and Electronics Engineers, 10667 Los Vaguenos Circle, Los Alamitos, CA 90720	(2)*
4.	*AI Expert*, Miller Freeman Publications, 500 Howard St., San Francisco, CA 94105	(2)*
5.	*Expert Systems - The International Journal of Knowledge Engineering*, Learned Information Limited, Besselsleigh Rd., Abingdon, Oxford, OX13 6LG, England.	(2)*

* Number of universities recommending the journal

5 COMPUTING ENVIRONMENT

A myriad of hardware and software currently exist in the expert systems market and the inventory is increasing as well as changing. While the competition in the computer industry is driving prices down, it has made the industry highly volatile. No one knows which of the companies will stay in the market for the next several years and what computing environment (hardware, software, operating system, networking mechanism, and peripherals) would be tomorrow's best. Given these price and technology uncertainties, choice of the computing environment for an expert systems course laboratory is a difficult one. In fact, it is prudent to include a topic on the survey of current environments, their distinguishing features, and the evaluation criteria in the expert systems course syllabus. The following sections describe the state-of-affairs based on the responses received in this monograph project, and also some thoughts on expert systems software selection criteria.

5.1 Expert System Software

A majority of the instructors teaching expert systems techniques have raised following questions.

(i) Should the course include programming in LISP, PROLOG or some other language?
(ii) Should the course laboratory work be based on an in-depth use of one expert system software or an overview of several shells?

Answers to the above questions can be sought in the current practices. Typically, wherever students have been given a choice, they have preferred PC-based tools. Therefore, most of the universities who offer a one-semester or one-quarter all-inclusive course have found it satisfactory to use one or two PC-based shells for use in class projects. They feel that high-end shells like KEE^{TM}, ART^{TM}, KNOWLEDGECRAFT^{TM}, OR NEXPERT OBJECT^{TM} will be difficult to learn and use in a one-semester course and should be acquired for more advanced courses and research. Another group of instructors feels that although PC-based shells do provide quick prototyping at affordable prices, they fail in supporting development of state-of-the-art expert systems. They suggest that an expert system software that supports hybrid programming environment like NEXPERT OBJECT^{TM} or KAPPA^{TM} are better suited for an expert systems course rather than those supporting only rule-based programming. At the very least, they suggest, the shell should support both rule-based and frame-based programming. Table 8 lists most used expert system software in courses currently being taught.

The price of expert system software is coming down and academic discounts up to 90% are available on most software. Workstation prices are also now affordable. It is now possible to purchase two high-end and four PC-based shells for less than

TABLE 1-7: Twenty Most Recommended Technical Papers

1. Bennett, J.S., and Engelmore, R.S., *SACON: A Knowledge-Based Consultant for Structural Analysis*, Stanford University, Dept. of Computer Science, Technical Report STANCS-78-699, September 1978. (2)*

2. Bobrow, D.G., and Stefik, M.J., *Perspectives on Artificial Intelligence Programming*, Science, Vol. 231, pp. 951-957, February, 1986. (3)*

3. Bobrow, D.G., Mittal, S., and Stefik, M.J., *Expert Systems: Perils and Promise*, Communications of the ACM, Vol. 29 (9), pp. 880-894, September 1986. (2)*

4. Buchanan, B., and Duda, R., *Principles of Rule-Based Expert Systems*, Stanford University, Dept. of Computer Science, HPP-82-14, pp. 24-55, August 1982. (2)*

5. Degroff, L., *Conventional Languages and Expert Systems*, AI Expert, pp. 32-26, April 1987. (2)*

6. Dreyfus, H., and Dreyfus, S., *Why Computers May Never Think Like People*, Technology Review, Edited at MIT, Vol 89 (1), pp. 42-61, January 1986. (2)*

7. Dym, C.L. *Expert Systems: New Tools for Computer-Aided Engineering*, Engineering with Computers, Vol. 1(1), pp. 9-25, April 1985. (2)*

8. Fikes, R., and Kehler, T., *The Role of Frame-Based Representation in Reasoning*, Communications of the ACM, Vol. 28(9), pp. 904-920, September 1985. (2)*

9. Genesereth, M.R., and Ginsberg, M.L., *Logic Programming*, Communications of the ACM, Vol. 28(9), pp. 933-941, September 1985. (2)*

10. Hart, A., *Knowledge Elicitation: Issues and Methods*, CAD, Vol. 17(9), pp. 455-462, November 1985. (2)*

11. Levitt, R., and Kunz, J., *Using Artificial Intelligence Techniques to Support Project Management*, Artificial Intelligence for Engineering Analysis, Design and Manufacturing, Vol. 1(1), pp.3-24, December 1987. (3)*

12. McDermott, J., *R1: A Rule Based Configurer of Computer Systems*, Artificial Intelligence Journal, Vol. 19(1), pp. 39-88, 1982. (2)*

13. Mittal, S., and Dym, C.L., *Knowledge Acquisition from Multiple Experts*, The AI Magazine, Vol. 6(2), pp. 32-36, Summer 1985. (4)*

14. Mittal, S., Dym, C.L., and Morjaria, M., *PRIDE: An Expert System for the Design of Paper Handling Systems*, IEEE Computer, Vol. 19(7), pp. 102-114, July 1986. (2)*

15. Nii, H.P., *Blackboard Systems: The Blackboard Model of Problem Solving and the Evolution of Blackboard Architectures (Part One)*, The AI Magazine, pp. 38-53, Summer 1986. (2)*

16. Nii, H.P., *Blackboard Systems: Blackboard Application Systems, Blackboard Systems from a Knowledge Engineering Perspective (Part Two)*, The AI Magazine, pp. 82-106, August 1986. (2)*

17. Sathi, A., Morton, T.E., and Roth, S.F., *Callisto: An Intelligent Project Management System*, The AI Magazine, pp. 34-52, Winter 1986. (2)*

18. Stefik, M.J., and Bobrow, D.G. *Object-Oriented Programming: Themes and Variations*, The AI Magazine, Vol. 6(4), pp. 40-62, Winter 1986. (4)*

19. Thompson, B., and Thompson, B., *Creating Expert Systems from Examples*, AI Expert, pp. 21-26, January 1987. (2)*

20. Wigan, M.R., *Engineering Tools for Building Knowledge Based Systems on Microcomputers*, Microcomputers in Civil Engineering, Elsevier, Vol. 1(1), 1986. (2)*

* Number of universities recommending the paper

TABLE 1-8: Ten Most Used Expert System Software for Class Projects

1.	INSIGHT2+™/LEVEL 5™	(7)*
2.	VP-EXPERT™	(4)*
3.	PERSONAL CONSULTANT PLUS™	(4)*
4.	OPS 5™	(3)*
5.	EXSYS™	(3)*
6.	DECIDING FACTOR™	(3)*
7.	KEE™	(2)*
8.	FRAMEKIT™	(2)*
9.	SMALLTALK™	(2)*
10.	M.1™	(2)*

* Number of universities that use the software

$15,000 and maintain their technical support services and updates in less than $5,000 per year. A library of about six software packages selected judiciously will provide adequate variations of knowledge representation schemes, inferencing mechanisms, integration capabilities and other problem solving comparisons to students, both for learning as well as for research.

5.2 Hardware for Expert Systems Laboratories

A wide range of hardware is currently being used at the civil engineering departmental labs for expert systems instruction. Observations of some of the course instructors on hardware selection are summarized in the following sections.

5.2.1 PC-Based Shells. PCs with 80286 processors, 80287 math coprocessors, 640K RAM, and 20 megabyte hard disks are adequate for most of the expert system shells. KAPPA™, however, required two megabyte of RAM; and some newer software like NEXPERT OBJECT™ required a 80386-based machine with nearly four megabytes of memory and an EGA card/monitor.

5.2.2 High-end Software. High-end shells with hybrid knowledge representation require a workstation with 12 to 20 megabyte of main memory, a minimum of 100 megabyte harddisk, more than four MIPS execution rate, UNIX operating system, a tape drive, and a 19 inch high resolution color graphics monitor.

5.2.3 An advanced classroom environment. A room capable of CRT projection with 6-8 workstations networked to a file server and connected to the campus EtherNet is needed to make this type of course more effective.

The number of computers dedicated to expert systems instruction is dependent on the target number of students in the course and the available resources. Texas A&M University has five PCs for its eight students, while the University of Washington has 40 IBM/AT computers for 25 students. Two to three students per computer can be satisfactorily scheduled and the existing departmental computing labs can be upgraded for teaching. However, if research is to be pursued, a separate lab for expert systems research and instruction would be a more desirable option.

6 DEVELOPMENT OF TEACHING EXPERTISE

At this time, the faculty trained to teach expert systems courses is very limited and it is unlikely that the number of civil engineering Ph.D.'s who have used expert systems techniques in their doctoral dissertation will match the demand, at least in the near future. Faculty for teaching expert systems techniques will, therefore, have to be developed through non-traditional ways. Some of the modes successfully used by various universities are cited below:

(i) Sending an interested faculty member for a semester to another university which offers courses on expert systems and other AI applications.

(ii) Encouraging faculty member(s) to take courses in AI in their computer science department, if available.

(iii) Acquiring two or three PC-based expert systems software and arranging in-house funds to support a graduate student and if possible faculty time for research that uses expert systems techniques.

(iv) Encouraging faculty members to team up with computer science faculty and submit research proposals relating expert systems with civil engineering problems, to various sponsors.

(v) Sending faculty member(s) out to take short courses offered by ASCE, various other organizations, and software vendors, then offering short courses for civil engineering students and, later, to the local industry.

(vi) Inviting visiting faculty from some other university for a semester to offer a course on expert systems and, thus, provide in-house training to the interested faculty.

(vii) Sending a faculty member out to work in an AI related industry on a continuing project.

A majority (at least 12 out of 15) of the universities currently teaching expert systems courses developed their faculty expertise using a subset of the above cited modes and further built on it through offering courses in their civil engineering departments. The keys to success have been the initiative and continual commitment of the faculty member and the administration's support in funding the purchase and make-up of the necessary hardware and software.

7 COURSE ORGANIZATION

The expert systems course is being offered as a technical elective in 13 out of 15 civil engineering departments. The primary focus of the course in almost all of the universities has been their graduate students. Seven of the 15 universities offer the expert systems course strictly for their graduate students while five of them open the course to the senior undergraduates.

7.1 Course Requirements

Most courses have required extensive reading of 1-2 text books, several chapters from 2-4 reference books, and 15-30 technical papers. Weekly or biweekly assignments are given on the class lectures and suggested readings. The suggested readings have also generally been included on the exams. The courses have also required the development of at least one prototype expert system, in most cases using PC-based expert system software. A few universities in Europe give exercises or projects in which students have to write simple programs using languages like LISP, PROLOG, or SMALLTALK™. Students find this assignment hard. Some of the instructors programmed a few examples using each of the expert systems software and made these available, greatly helping the students in the learning process.

7.1.1 Class Projects. All of the course instructors have recognized that the best method of teaching the various knowledge engineering concepts is by enabling the students to see them in an application of their interest. Consequently, they have required the development of a prototype expert system, which has been weighted heavily (40% in a majority of cases) in the evaluation of the student's grade in the course. Although small groups (<3) are acceptable to some of the universities, students have mostly worked individually on these projects. A good number of these prototypes have been published in technical journals and a few of them have progressed into graduate research. In fact, use of the class project for research development has been encouraged by most of the instructors. Some of the universities have required their students to submit a term paper and a user manual on their system, while some others have required a class presentation during the final week.

For monitoring the lab work, most of the universities have broken the development of the class project into several modules and they require their students to submit each module in sequence and grade each module. These projects have typically included the following modules:

1. Choice of a domain and narrative description of the proposed system.
2. Learning an expert system shell.
3. Formal description of the proposed system, sample rules.
4. Identification of domain experts and other knowledge sources; Knowledge acquisition.
5. Sample implementation.
6. Building, testing and documentation, final project report.
7. Class presentation.

One notable aspect in almost all of the class projects has been the lack of emphasis on knowledge acquisition from human experts, the only exception being the National School of Roads and Bridges in France.

7.1.2 Grading Policies. The course grading has been based on: (i) home assignments, (ii) exams, and (iii) class project. Generally, five to ten home assignments have been given on class lectures and suggested readings, determining 20% to 40% of the course grade. About half of the universities give mid-term and/or final exams, assigning 10% to 40% weight. All of the universities assign about 40% weight to the class project(s), which includes the development of a prototype expert system. A typical weight distribution for student evaluation can be given as below:

Home Assignments	30%
Exams (Midterm and Final)	30%
Project(s)	40%

7.2 Course Format

7.2.1 Course Delivery Schedule. Currently, most courses are being taught as a one-semester or one-quarter course covering a broad spectrum of material on expert systems techniques. A typical delivery schedule has been two 75 to 80 minute lectures and a tutored lab session per week. One university schedules three one-hour lectures per week without any lab sessions. Students devote several more hours per week in computer labs on their project.

7.2.2 Class Size. The class size has ranged from 8 to 40, with a typical size of 25 students. Besides the sizes of the programs within civil engineering, the factors determining class size have been: (i) the availability of computers and expert system software and (ii) the instructor's time. Some of the universities employ teaching assistants to help the students on their projects and home assignments, thus affording larger class sizes.

7.2.3 Prerequisites. At this time, very few universities have formal prerequisites for registering in their expert systems courses, assuming that the students desirous of taking this senior level and/or graduate course would have had adequate familiarity with personal computing. The few that have prerequisites, require a course in computer science or AI. Some of the instructors who have taught expert systems for a couple of years have commented that those of the students who had conventional programming skills in procedural languages such as FORTRAN, Pascal, or C did a significantly better job on their projects.

7.2.4 Guest Speakers. About one-third of the universities currently teaching expert systems courses have invited guest speakers into the classrooms to share their experiences in building and using knowledge based expert systems. These speakers have come either from industry, from their computer science department, or from within the civil engineering department from areas other than the instructor's. Speakers from industry, wherever available, have been very interesting and motivating to students as they present the total process of the development of a KBES. Hearing first hand on various issues -- what tools he used, how he acquired knowledge from human experts and other sources, and how useful the industry finds these expert systems -- has been very instructive.

8 SHORT COURSES

Short courses on expert systems are typically taught to industry users in a period of three to ten days and are, thus, more demanding on the instructor who has to condense the total material, and lecture for very long hours at a stretch. Then, the audience is very diverse in interest as well as capabilities. Their interest lies in learning applications that have immediate impact on their current work. Theoretical concepts and formulations are not emphasized. One positive aspect of the short course is that most of the participants are experienced professionals, some of them may serve as domain experts in building prototype expert systems for class projects. Their participation may also make the course instructors aware of what industry is looking for.

At this time, most of the short courses are offered by the expert system software manufacturers whose primary focus is to teach the use of their product. These courses which are given in 1-5 days have a narrow basis from expert systems education viewpoint. However, all of these courses include hands-on training workshops and may prove fruitful for beginners. Some AI related engineering consultants also offer short courses that include workshops. A few companies have also developed expert systems short courses for in-house training of their staff.

Very few civil engineering departments are currently offering short courses on expert systems. The American Society of Civil Engineers (ASCE) has been teaching short courses since 1985, using experienced faculty members from universities.

This course is described in Chapter 11 of this monograph. One university in the U.K. teaches a short course on expert systems to its senior civil engineering undergraduate students. One university in the U.S.A. teaches a one-week summer course on Knowledge-based Expert Systems. Another university in the U.S.A. has developed a short course for the Boeing Aerospace Company. This course has also been taught in several universities in Europe and is described in Chapter 10 of this monograph.

9 COURSE ASSESSMENT

In all of the civil engineering departments where courses in expert systems techniques have been offered, students have shown a keen interest in the technology and willingly devoted more time and effort than other courses with equivalent credit hours. Course enrollments have increased in subsequent offerings and a good percentage of graduate students have used expert system techniques in their theses.

Many instructors have reported that, in spite of more than usual efforts put in by them, the results have been very satisfying -- both to them as well as to the students. During the process of building their prototype, students learn to understand and appreciate the value of professional experience and expertise in the civil engineering world.

Steve Fenves of CMU, in one of his letters to the author describing his experience of teaching an expert systems course, wrote:

> Concerning my assessment, I think that the course does achieve its objective: it de-mystifies the topic of expert systems; demonstrates to the students that the ES methodology is just another set of tools in the toolkit of a computer-oriented civil engineer; and that the "traditional" division of "expert" vs. "knowledge engineer" is unnecessary. ... The negative factors are that students consistently underestimate the difficulty of acquiring domain knowledge, and that students well drilled in structural procedural program development have to be "untaught" to accept declarative knowledge representation.

10 CONCLUDING REMARKS

Teaching an expert systems course has not been easy, according to the course instructors. A course of this type requires a lot of planning: the selection, acquisition, and management of computing environment; lecture topics; and domain-specific problems. Ideally, the instructor must have a thorough knowledge of the capabilities and limitations of the shell(s) to be used. In a fast changing computing environment, this is not an easy task. In most of the civil engineering departments the expert systems course did not start as a part of their academic curriculum but as a 'special topic', purely on the initiative and enthusiasm of one faculty. In one case, when the instructor moved to another university, the course was not continued.

As of today, there is relatively little experience with teaching expert systems courses to civil engineering students. This lack is understandable because the subject is new. Some of the obvious problems that will have to be solved are:

(i) Development of faculty interest/expertise in the area.
(ii) Establishment of Applied AI computing laboratories.
(iii) Graduate research using expert system techniques.
(iv) Structure of the course
(v) Production of text books
(vi) A balance between the various civil engineering areas: construction, environmental, geotechnical, structures, and transportation.

It is hoped that this monograph will help in resolving some of the above problems. The primary onus, however, lies on the faculty who recognize the considerable potential of expert systems technology.

11 FUTURE DIRECTIONS

For a thorough coverage of the technology so that civil engineers, after graduation, feel comfortable in using expert systems techniques in their work, one single course is not enough. Several instructors who have taught expert systems courses feel that the material should be covered in a series of two courses. The first course should include AI concepts; programming using one of the AI languages: PROLOG or LISP; and knowledge of a few PC-based shells like LEVEL 5™, PERSONAL CONSULTANT PLUS™, VP-EXPERT™, or KAPPA™. The second course should include several lectures and practice sessions on knowledge acquisition methods. This very important topic of knowledge acquisition is not covered in detail in any of the courses, currently. The second course should require the development of an expert system using a high-end shell like KEE™, ART™, KNOWLEDGECRAFT™ or NEXPERT OBJECT™. Those students who have experience with programming languages like C, Pascal, PROLOG, or LISP should be able to write their expert systems in the language of their interest. This system should solicit knowledge from human experts and other sources, should integrate with algorithmic programs, databases, and graphic packages; and should be close to an operational prototype. One university in France offers a sequence of two courses close to the above suggested format. In the second course, students work on real problems in which they develop operational expert systems using knowledge from human experts.

Acknowledgements. Most of the material presented in this chapter has come from the information supplied by many universities in the U.S.A. and abroad in response to our survey. The author wishes to acknowledge the contributions of the survey respondents. Some of the data has been taken from several of the chapters of this monograph. The author acknowledges the authors who have participated in this monograph.

12 REFERENCES

Brown, David C., *A Graduate-Level Expert Systems Course*, AI Magazine, Fall 1987.

Maher, M.L., (editor), *Expert Systems for Civil Engineers: Technology and Application*, American Society of Civil Engineers, New York, 1987.

**TECHNICAL COUNCIL ON
COMPUTER PRACTICES**
COMMITTEE ON EXPERT SYSTEMS

Appendix

**Expert Systems Education in Civil Engineering -
<u>Survey of the State-of-the-Art Teaching Practices</u>**

(1) Is your institution offering course(s) in Expert Systems in the Department of Civil Engineering?

   [ ]   offering a graduate/undergraduate course at the present time
   [ ]   planning to start offering a course in the near future
   [ ]   has no plans, but feel such a course, if offered, would be useful
   [ ]   feel such a course is not necessary in a civil engineering curriculum

(2) If you are currently offering a course in expert systems or are planning to offer one in the near future, please supply the following material:

   (i)   Course Objectives
   (ii)   Course Outline
   (iii) Text book(s) used
   (iv)   Suggested Readings (please attach a copy of any class handouts)
   (v)   Sample problems, assignments, projects and exams
   (vi)   Prerequisites
   (vii) Any other material considered relevant
   (viii) Your assessment of the course

(3) Are any members of your institution involved in expert systems research? If yes, please state their names, addresses and interest areas.

   _____   _____
   _____   _____
   _____   _____

(4) Would you be interested in authoring a paper on describing your experiences in teaching expert systems techniques?

(5) Comments if any _____

Address reply to:

Satish Mohan
Department of Civil Engineering
State Univ. of New York at Buffalo
Buffalo, NY 14260, USA

AMERICAN SOCIETY OF CIVIL ENGINEERS
345 East 47th Street
New York, New York 10017-2398   (212) 705-7496

Chapter 2
Teaching Expert Systems Techniques at Wayne State University

by Tomasz Arciszewski, A.M. ASCE

1 INTRODUCTION

Research on expert systems techniques was initiated in the Civil Engineering Department at Wayne State University in 1985. At this time the author began working with Dr. Wojciech Ziarko of the Computer Science Department, University of Regina, Canada, on the application of inductive learning in engineering. This work was directly related to earlier cooperation on innovative design methodology, particularly on the development of computer tools supporting morphological analysis. For this reason the research had a very strong methodological flavor. Its objectives were twofold: (i) to investigate the feasibility of applying inductive systems in engineering and, (ii) in the case of promising results, to develop a methodology of use for such systems for different engineering purposes. The research was conducted using several experimental inductive systems based on the theory of rough sets developed at the University of Regina. The initial results were quite promising and caught the attention of a number of Ph.D. students, who became interested in inductive learning for civil engineering applications, and in expert systems techniques in general. This growing student research activity was supported only by individual instruction and selected recommended readings. This was obviously insufficient and a need for more systematic and comprehensive expert systems techniques training was noted. Because of the size of the Civil Engineering Department and existing teaching commitments, it was found impossible for the author to introduce another full course. To solve this antinomy between needs and available resources, an innovative solution was sought and found in the form of "distributed expert systems techniques education." This is education distributed between two different graduate courses, usually taught by the author, which together provide a basic understanding of expert systems techniques in civil engineering. Each course includes several lectures related to expert systems, and together provide the most basic understanding of expert systems techniques and their potential in civil engineering. For the first time, several lectures on the subject were included in course CE740 - Optimal Structural Design, taught in the Winter term, 1986. This was continued in CE701 - Civil Engineering Decision Processes, offered in the Winter term of 1987. The sequence was continued in the Fall of 1988, when again CE740 was offered, although this time the discussion of expert systems techniques was reduced to the applications of inductive learning in structural engineering only. Enrollments in CE740 in 1986 and 1988 were nine and five students respectively, while fourteen students enrolled in CE701. The latter course was also attended by a faculty member and two Ph.D. students from the Department of History, who were interested in possible applications of expert systems in their historical research. There were no special prerequisites related to expert systems and the coverage of this area was intended to be comprehensive and self-contained.

This experimental distributed teaching of expert systems techniques was found sufficient to provide an initial understanding of this area and to raise the students' interests, but was definitely insufficient to prepare students for research activity in this area. The author came to the conclusion that a complete 600-level graduate course available to both seniors and graduate students is a must. Such a course will be offered in the Fall term of the 1990/91 academic year.

2 COURSE OBJECTIVES

The objective of distributed expert systems education is to provide graduate students with a basic understanding of expert systems and their potential in civil engineering. Considering the very limited coverage of the subject, the building of practical skills in the area of development of expert systems and knowledge acquisition through computer learning was of secondary importance.

This course was based on the assumption that expert systems techniques should be taught in the context of design methodology and engineering decision making. It was assumed that an expert system is a new engineering tool supporting decision making. Its user should understand both the internal workings of an expert system and the use of this system in the process of engineering design or knowledge acquisition.

This understanding should help him/her to recognize the importance of expert systems techniques in the context of changes in engineering and computer science, and to identify their potential applications in civil engineering. This knowledge should also raise the student's natural curiosity and encourage him/her to continue in the area of expert systems techniques.

3 COURSES SYLLABI

This section covers three different courses, which will be discussed separately. The syllabi listed are related only to expert systems techniques as they were taught in these three courses.

3.1. CE740 - Optimal Structural Design. Offered in the Winter of 1987.

1. Introduction: Artificial Intelligence. (One lecture)

Definitions of human and artificial intelligence. Roots of artificial intelligence, including cognitive science and formal logic. Areas of AI research. Symbolic programming languages, including LISP and Prolog.

2. Expert Systems. (One lecture)

History of research and implementation. Different definitions. Comparison with database programs. Components of an expert system. Different forms of knowledge representation, including semantic networks, object-attribute-value triples, rules, frames, logical expressions, and models. Different forms of reasoning, including backward and forward chaining.

3. Knowledge Engineering and Development of Expert Systems. (One lecture)

Roots of knowledge engineering, including heuristics, cognitive science, computer science, design methodology, and AI. Development of expert systems: direct (using symbolic programming languages) and indirect (using development shells). Comparison of several available development shells.

4. Demonstration of a Symbolic Language and a Development Shell. (One lecture)

The demonstration utilized a computer program for a structural purposes written in GC LISP, and a Texas Instruments Consultant Plus expert systems development shell.

This course included a student project: development of a simple expert system using a shell, as described in Section 7.

3.2. CE701 - Civil Engineering Decision Processes. Offered in the Winter of 1988.

1, 2, 3 - as in CE740, Winter 1986, (two lectures)

4. Expert System as Decision Support Tool. (Half lecture)

Decision making process. Algorithm as a sequence of decision rules. Decision rules and their importance. The expert system as an equivalent of decision-related knowledge in the form of a system of decision rules.

5. Computer Learning from Examples. (One lecture)

Basic definitions and types of inductive learning, including learning based on the probabilistic, Darwinian, and rough sets approaches. Available inductive systems, including Auto Intelligence, BEAGLE, Super Expert, and several experimental systems based on the theory of rough sets.

6. Inductive Learning Process: User's Perspective. (One and a half lecture)

Basic models of the inductive learning process, including point, linear, and mixed models. Monitoring of an inductive learning process: inductive learning process spreadsheet and decision rules network. Selection of examples, including purely random, coincidental, balanced, and corrective selection. Global and local control criteria.

3.3. CE740 - Structural Optimal Design. Offered in the Fall of 1988.

1. Application of Inductive Systems in Knowledge Acquisition. (One lecture)

Extraction from examples of a system of decision rules governing the problem of classification of wind bracings in skeleton structures of a tall building. Demonstration of an experimental inductive system, ANLYST, based on the theory of rough sets.

This course included a student project in the area of knowledge acquisition, as described in Section 7.

4 TEXTS AND REFERENCES

When distributed teaching of expert systems techniques was initiated in 1986, no adequate textbook was available. Therefore a number of different books and research papers were used as references for individual parts of the teaching program. These references are given below.

1. Mishkoff, H.C., Understanding Artificial Intelligence, Howard W. Sams & Co., 1985.

2. King, D., Harmon, P., Expert Systems, John Wiley and Sons, 1985.

3. Arciszewski, T., Mustafa, M. and Ziarko, W., "A Methodology of Design Knowledge Acquisition for Use in Learning Expert Systems," International Journal of Man-Machine Studies, No.27, 1987.

4. Arciszewski, T., Mustafa M., "Inductive Learning Process: The User's Perspective," invited chapter, in the book Machine Learning, edited by R. Forsyth, Chapman and Hall, 1989.

5. Maher, M.L., (Editor), Expert Systems for Civil Engineers, American Society of Civil Engineers, 1987.

6. Arciszewski, T., "Design Making Parameters and Their Computer-Aided Analysis for Wind Bracings," Advances in Tall Buildings, Van Nostrand Publishing Company, 1985.

7. Arciszewski, T., Ziarko W., "Adaptive Expert System for Preliminary Engineering Design," Revue Internationale D.E. CFAO et D'Intographie, Vol. 2, No. 1, 1987.

8. Arciszewski, T., Mustafa, M. and Ziarko, W., "A Methodology of Design Knowledge Acquisition for Use in Learning Expert Systems," International Journal of Man-Machine Studies, No.27, 1987.

9. Arciszewski, T. and Ziarko, "Adaptive Expert System for Preliminary Design of Wind Bracings," Second Century of Skyscrapers, Van Nostrand Publishing Company, 1988.

10. Arciszewski, T., Mustafa M., "Inductive Learning Process: The User's Perspective," invited chapter, in the book Machine Learning, edited by R. Forsyth, Chapman and Hall, 1989.

5 COURSE ORGANIZATION

The entire program of distributed expert systems education included nine lectures spread over a period of three years. These lectures were supposed to be taken by a group of students interested in the design methodology and civil engineering applications of expert systems. The sequence of lectures and laboratory projects reflected this goal. Three clusters of lectures and two home assignments supplemented classroom and lab work.
The individual clusters of lectures could be called:
 1. Introduction to Artificial Intelligence and Expert Systems
 2. Methodology of Inductive Learning
 3. Application of Inductive Systems in Structural Knowledge Acquisition.

Laboratory projects were related to the clusters of lectures. The first project required the development of an expert system using a shell, while the second required the use of an inductive system in the process of knowledge acquisition.

Distributed expert systems education has been prepared and implemented as a temporary measure. This innovative solution was used in a situation requiring immediate training in this area for highly qualified and motivated graduate students at a time when the existing system of courses did not permit the immediate introduction of a special course, or a sequence of courses entirely devoted to AI and expert systems in civil engineering applications. Our solution enabled us to gain teaching experience and to test the students' response to a body of knowledge much different than traditional analytical or descriptive structural knowledge.

The initial student response was one of surprise. It was the first time they had encountered a different approach to problem solving and dealing with engineering knowledge. However, they quickly became involved in the lectures and found many potential applications for expert systems. In particular, part-time graduate students, with significant engineering experience, were able to relate expert systems techniques to their practice, and their comments were usually very interesting and inspiring. Students were also surprised that inductive learning, still essentially the domain of computer scientists, is presently available to engineers, and that inductive systems can be used for practical engineering purposes with very limited training. In general, the students' response to distributed expert systems education was positive, and the author has been strongly encouraged to develop a more complete educational program in this area, based initially on a single graduate course, Expert Systems in Civil Engineering.

6 HARDWARE AND SOFTWARE USED

All three courses described in this chapter were conducted using the resources of the Intelligent Computers Center, established at Wayne State's Civil Engineering Department in 1986. The mission of the Center is to conduct integrated research on design methodology, expert systems and inductive learning in engineering. The Center features a network of computers composed of a Texas Instruments Explorer LX System and four Texas Instruments Business Professional Computer Workstations. The Center possesses a number of expert systems development shells, including Texas Instruments Personal Consultant Plus, several experimental inductive systems based on the theory of rough sets, and commercial inductive systems including BEAGLE, Super Expert, Auto Intelligence, and 1st. Class.

A Texas Instruments PC+ expert system development shell was used for the demonstration of expert system development principles and in student projects related to the building of an expert system. In the area of knowledge acquisition, two different inductive systems were used: an experimental system, ANLYST, based on the theory of rough sets and developed by Voytech Systems, Inc., and Super Expert, developed by Softsync, Inc.

The available computer hardware and software was found sufficient for the limited scope of this course. It is obvious, however, that more elaborate training in this area will require a dedicated computer laboratory with at least twelve computer workstations and one symbolic computer. IBM AT or PS-2 computers could be used with a Gold Hill LISP programming environment and Texas Instruments PC+ expert systems development shells. A Texas Instruments Explorer seems to be a natural choice as a symbolic computer, which is necessary for research and advanced development purposes. Another interesting possibility is to use fifteen Apple Mac II computers with the software described above, and to equip several computers, (20 to 25%) with Texas Instruments Micro Explorer boards. Micro Explorer has sufficient performance for most research and development purposes in the area of the artificial intelligence, is relatively inexpensive, and is much more user-friendly than the Explorer itself. Both possibilities have a number of advantages and disadvantages and the final choice of the hardware and software configuration should be based not only on our needs in the area of expert systems education, but also on criteria related to other uses for a given computer network.

7 LEVEL OF STUDENT INVOLVEMENT

There were two laboratory projects, each requiring the use of different expert systems techniques and computer tools.

The first project was to develop an expert system for the selection of the proper type of washer to be used in structural joints utilizing ASTM A325 or A490 bolts in accordance with the Manual of Steel Construction, Load and Resistance Factor Design. Several different types of washers had to be considered, including standard, hardened, hardened beveled, and plate washers. The selection of the type of washer is based on the character of the connection under consideration, and the RCSC Specification for Structural Joints gives eight complex requirements for washers which have to be satisfied. The expert system was to be developed using the Texas Instruments PC+ development tool.

Students were given a demonstration on the use of the PC+, and a Graduate Teaching Assistant with some experience in the use of this shell was assigned to provide additional advice and support.

Students had problems with the proper understanding of the washer requirements, with their formulation in the form of if-then rules, and with the computer implementation of these rules in an expert system. It was found that the use of TI PC+ required much more instruction and experience than initially anticipated. Not all students were able to build complete expert system prototypes, although all thought that it was a worthwhile experience, and their work with PC+ gave them initial experience and a better understanding of expert systems techniques.

The second project was in the area of knowledge acquisition. It included the use of two different inductive systems for the extraction of decision rules from a given body of examples. It had a much more unique character than the first project and therefore it will be discussed here in some detail, including the project description as it was given to students. Students also received packages containing copies of four publications related to the subject of their project.

Project Description:

A computer program has been developed for the nondeterministic optimization of wind bracings in skeleton structures of tall buildings. This program generates new types of wind bracings. The generated types have to be classified before they are evaluated by a group of experts. The classification of thousands of generated types is very time-consuming and cannot be completed within the imposed time limit. To meet the deadline the use of inductive learning was proposed: an inductive system will be used to extract classification rules from examples evaluated by experts, and these rules will be used later to classify all generated types of wind bracings. You are an expert in the areas of structural optimization and tall buildings. You are requested to conduct the following tasks:

1. Prepare 97 examples representing known types of wind bracings classified as truss, frame, and truss-frame bracings.
2. Use an inductive system to extract classification rules. At least three different inductive learning sequences should be used.
3. Analyze classification rules extracted at individual stages of the inductive learning process and determine their structural character, generality and validity.
4. Prepare your general comments regarding the feasibility of inductive learning in the determination of classification rules in structural engineering.

The preparation of examples of different types of wind bracings turned out to be quite difficult for students. The development of these examples was only preparatory with respect to the inductive learning experiment. Therefore students were given additional instruction in the typology of wind bracings in skeleton structures in accordance to reference 6 and were provided with 108 examples of types of wind bracings, divided into three families, and classified as frame, truss, and truss-frame bracings.

TEACHING EXPERT SYSTEMS

Parameters	Feasible States			
Number	1	2	3	4
1. Static character of joints	rigid	hinged	rigid & hinged	
2. Number of bays entirely occupied by bracing	1	2	3	
3. Number of vertical trusses	0	1	2	3
4. Number of horizontal trusses	0	1	2	3
5. Number of horizontal truss systems	0	1	2	3
6. Material of core used	0	steel	reinforced concrete	
7. Number of cores	0	1	2	3
8. Structural character of external elements	0	columns & beams	cables	bar discs
9. Static character of bottom external joints	0	rigid	hinged	rigid & hinged

Table 2-1. Typologic table for wind bracings

All examples of known types of wind bracings were prepared using the typologic table (Table 2-1) taken from reference 6. A type of wind bracing is understood here as a compatible combination of feasible states taken from a typologic table, when for each variable from the table one feasible state is taken at a time (reference 6).

Three groups of examples of wind bracing types were prepared for individual families of wind bracings. For instance, a family of frame bracings is shown in Fig. 2-1 (reference 6) in graphical form, and in the form of examples in Fig. 2-2. All examples of different types of wind bracings considered in the experiment are shown in Table 2-2.

Students were encouraged to be innovative and to select an inductive system, or systems, from five made available for this project, including four different experimental systems based on the theory of rough sets, and a commercial system, Super Expert. Since they were also requested to use three different inductive learning sequences, the total number of expected experiments was between three and fifteen. Among the group of five students working on this project two subgroups could be distinguished: three students without any previous exposure to inductive learning, and two students with some experience in this area. This division could also be noted in the students' reports.

Those students without experience conducted inductive experiments as instructed, while the two more experienced students

EXAMPLE NUMBER	1	2	3	4	5	6	7	8	9	CLASSIFICATION
1	1	1	1	1	1	1	1	2	3	FRAME-BRACING
2	3	1	1	2	1	1	1	2	3	-
3	3	1	1	3	1	1	1	2	3	-
4	3	1	1	4	1	1	1	2	3	-
5	3	1	1	1	2	1	1	2	3	-
6	3	1	1	1	3	1	1	2	3	-
7	3	1	1	1	4	1	1	2	3	-
8	1	1	1	1	1	2	2	2	3	-
9	1	2	1	1	1	1	1	2	2	-
10	3	2	1	2	1	1	1	2	2	-
11	3	2	1	3	1	1	1	2	2	-
12	3	2	1	4	1	1	1	2	2	-
13	3	2	1	1	2	1	1	2	2	-
14	3	2	1	1	3	1	1	2	2	-
15	3	2	1	1	4	1	1	2	2	-
16	1	2	1	1	1	2	3	4	2	-
17	1	3	1	1	1	2	4	4	2	-
18	1	3	1	1	1	1	1	2	2	-
19	3	3	1	2	1	1	1	2	2	-
20	3	3	1	3	1	1	1	2	2	-
21	3	3	1	4	1	1	1	2	2	-
22	3	3	1	1	2	1	1	2	2	-
23	3	3	1	1	3	1	1	2	2	-
24	3	3	1	1	4	1	1	2	2	-
25	1	3	1	1	1	2	2	4	2	-
26	1	3	1	1	1	2	3	4	2	-
27	2	1	2	1	1	1	1	2	3	TRUSS-BRACING
28	2	1	2	2	1	1	1	2	3	-
29	2	1	2	3	1	1	1	2	3	-
30	2	1	2	4	1	1	1	2	3	-
31	2	1	2	1	2	1	1	2	3	-
32	2	1	2	1	3	1	1	2	3	-
33	2	1	2	1	4	1	1	2	3	-
34	2	1	2	1	1	2	2	2	3	-
35	2	2	3	1	1	1	1	2	3	-
36	2	2	3	2	1	1	1	2	3	-
37	2	2	3	3	1	1	1	2	3	-
38	2	2	3	4	1	1	1	2	3	-
39	2	2	3	1	2	1	1	2	3	-
40	2	2	3	1	3	1	1	2	3	-
41	2	2	3	1	4	1	1	2	3	-
42	2	2	3	1	1	2	3	4	3	-
43	2	3	4	1	1	2	4	4	3	-
44	2	3	4	1	1	1	1	2	3	-
45	2	3	4	1	2	1	1	2	3	-
46	2	3	4	1	3	1	1	2	3	-
47	2	3	4	1	4	1	1	2	3	-
48	2	3	4	1	1	2	2	4	3	-
49	2	3	4	1	1	2	3	4	3	-
50	3	1	2	1	1	1	1	2	3	FRAME-TRUSS BR'G
51	3	3	3	1	1	1	1	2	3	-
52	3	3	3	2	1	1	1	2	3	-
53	3	3	3	3	1	1	1	2	3	-
54	3	3	3	4	1	1	1	2	3	-
55	3	3	3	1	2	1	1	2	3	-
56	3	3	3	1	3	1	1	2	3	-
57	3	3	3	1	4	1	1	2	3	-
58	3	1	2	2	1	1	1	2	3	-
59	3	1	2	3	1	1	1	2	3	-
60	3	1	2	4	1	1	1	2	3	-
61	3	1	2	1	2	1	1	2	3	-
62	3	1	2	1	3	1	1	2	3	-
63	3	1	2	1	4	1	1	2	3	-
64	3	1	2	1	1	2	2	2	3	-
65	3	3	4	1	1	1	1	2	3	-
66	3	3	4	1	2	1	1	2	3	-
67	3	3	4	1	3	1	1	2	3	-
68	3	3	4	1	4	1	1	2	3	-
69	3	3	2	1	1	1	1	2	2	-
70	3	3	2	2	1	1	1	2	2	-
71	3	3	2	3	1	1	1	2	2	-
72	3	3	2	4	1	1	1	2	2	-
73	3	3	2	1	2	1	1	2	2	-
74	3	3	2	1	3	1	1	2	2	-
75	3	3	2	1	4	1	1	2	2	-
76	3	2	3	1	1	1	1	2	4	-
77	3	2	3	2	1	1	1	2	4	-
78	3	2	3	3	1	1	1	2	4	-
79	3	2	3	4	1	1	1	2	4	-
80	3	2	3	1	2	1	1	2	4	-
81	3	2	3	1	3	1	1	2	4	-
82	3	2	3	1	4	1	1	2	4	-
83	3	2	3	1	1	2	3	4	4	-
84	3	3	4	1	1	2	4	4	4	-
85	3	3	4	1	1	1	1	2	4	-
86	3	3	4	1	2	1	1	2	4	-
87	3	3	4	1	3	1	1	2	4	-
88	3	3	4	1	4	1	1	2	4	-
89	3	3	3	1	1	1	1	2	4	-
90	3	3	3	2	1	1	1	2	4	-
91	3	3	3	3	1	1	1	2	4	-
92	3	3	3	4	1	1	1	2	4	-
93	3	3	3	1	2	1	1	2	4	-
94	3	3	3	1	3	1	1	2	4	-
95	3	3	3	1	4	1	1	2	4	-
96	3	3	4	1	1	2	2	4	4	-
97	3	3	4	1	1	2	3	4	4	-

Table 2-2. Examples of different types of wind bracings

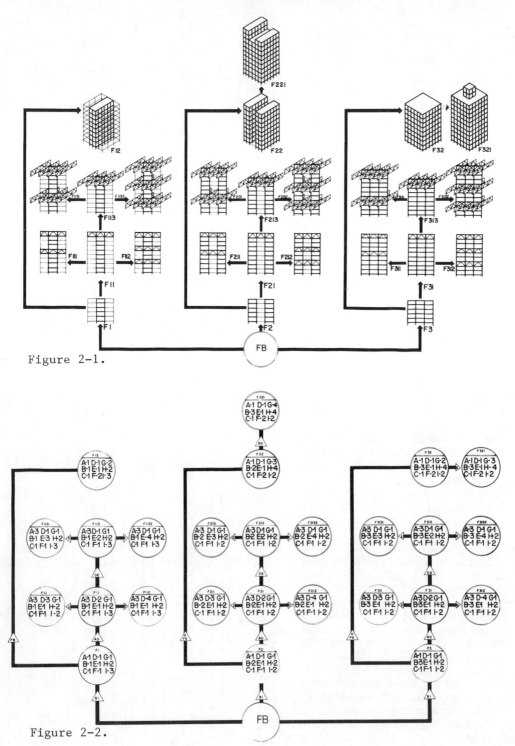

Figure 2-1.

Figure 2-2.

considered this project as a challenge. They were trying to use available inductive systems taking advantage of their experience, including additional qualitative analysis of examples and variables (data preprocessing) and a comprehensive comparison of their results. It should be noted, however, that all students found our very limited instruction and package of references sufficient for the purposes of their projects, and all students obtained reasonable results.Results obtained by one of the students (Ramzi Elachkar) using a three-stage learning process and an inductive system based on the rough sets theory and probabilistic learning algorithm implemented for a single decision tree will be discussed here. In this case 30 examples were used in the first stage, and 60 and 97 in the second and third, respectively.The first stage produced only five decision rules, using two condition variables. The student considered them correct, but rather obvious:

1. If joints are rigid, then a wind bracing is of a frame type.
2. If joints are hinged, then a wind bracing is of a truss type.
3. If joints are rigid and hinged and there are no vertical trusses, then this bracing is of a frame type.
4. If joints are rigid and hinged and there is a single vertical truss, then this bracing is a truss-frame type.
5. If joints are rigid and hinged and there are two vertical trusses, then this bracing is a truss-frame type.

The second stage resulted in eight decision rules, more specific and based on three condition variables. For example, rule No. 1 from the first stage was developed into two equivalent rules:

1. If joints are rigid and one bay is occupied by a bracing, then this bracing is of a frame type.
2. If joints are rigid and two bays are occupied by a bracing, then this bracing is of a frame type.

It is interesting to note that one rule was correct only in the case of 30 examples, and this rule was eliminated in the third stage. This rule was as follows:

If three bays are occupied by a bracing, then this bracing is of a truss-frame type.

The third stage produced thirty decision rules, based on two, three, four, and five condition variables. There are eight decision rules for the identification of frame bracings of different generality. The most general is as follows:

If joints are rigid and the material used for the core is steel, and if external structural members are in the form of columns and beams, then this bracing is of frame type.

The student correctly observed that "When we did not have enough examples, strange rules were generated...." and "An inductive learning system should use as many examples as possible....", which is correct only in the case of this experiment, when the learning process has not been completed and more examples could be effectively used.

Another student (Khaldoum Mhaimeed) began his work using a single-stage learning process. He also selected an inductive system based on the theory of rough sets and a probabilistic learning algorithm implemented for a single decision tree. He obtained the same system of thirty decision rules extracted in the three-stage learning process described above. Next he performed a multistage learning process and remarked that "I prefer the uniform model because it provides a good insight into the learning process and provides a better understanding of the problem."

One of the "experienced" students, Helmuth Koller, preprocessed the given variables describing wind bracings to make his inductive learning shorter and to obtain more general decision rules. He used a special system developed for this purpose, which is a part of all inductive systems based on the theory of rough sets. This system indicates which variables are necessary to conduct the extraction of decision rules from a given body of examples, considering only variable occurring in the final decision rules with at least three different values. The result of the preprocessing was quite surprising. Only two variables were selected: the static character of joints, and the number of vertical trusses. These variables were used to extract decision rules from examples, and only six such rules, based on one or two condition variables, were generated for the classification of all three types of wind bracings.

Another "experienced" student, Hinurimawan Sudarbo, compared Super Expert with four different systems based on the theory of rough sets. In all experiments he used a three-stage learning process. He observed that different rough sets-based systems produce decision rules of different complexity, and that one system, based on the significance factor approach, produces results identical to those obtained using Super Expert. Other systems produce larger numbers of decision rules, which are also of a much more detailed character and are based on up to seven variables, versus two in the rules produced by Super Expert.

The students produced non-trivial results and their work with different inductive systems definitely improved their understanding of inductive learning potential in civil engineering applications.

8 COURSES CRITIQUE

The distributed expert systems techniques education described in this chapter was prepared and implemented by the author only as a temporary measure to fill a gap in the existing civil engineering offerings at Wayne State University. The initial goals were reached: a number of graduate students were exposed to expert systems techniques, a lot of interest in this area was generated, we stimulated the interest of other faculty in expert systems, and we made possible the introduction of a regular course on expert systems. This course will be offered in the fall term of 1989. It should be noted, however, that the distributed teaching described in this chapter is not sufficient to train graduate students interested in research on expert systems applications in civil engineering. It should be considered only as an introduction to regular expert systems courses.

9 FUTURE PLANS

The teaching experience gained from distributed expert systems techniques education, will be used in the preparation and teaching of a regular course on expert systems, as mentioned the previous in section. This single course, although it will represent an improvement, will still be insufficient to provide a well-balanced education in expert systems in civil engineering, particularly considering the progress in this area and the research results becoming available every day. It seems justified to predict a need at Wayne State University for an entire graduate program in artificial intelligence in engineering, which would be run jointly by computer scientists, industrial engineers and experts on the methodology of expert systems development and computer learning. This program should include a cluster of about ten graduate courses specifically developed to train experts in the area of engineering applications of artificial intelligence. It should be implemented during the next two to three years, to keep up the present momentum and to meet the growing needs and interest in the applications of artificial intelligence in engineering.

Chapter 3
Teaching Expert Systems Techniques At Louisiana State University

by Demetre P. Argialas, A. M. ASCE

1 INTRODUCTION

Current and potential developments in expert system applications appear to have the capability to affect industry, research, and education in a substantial manner. The practicing engineer will be impacted by expert systems techniques because of their potential for enhancements in productivity by (1) making expertise available to novices, (2) making expertise available when needed, (3) improving the quality and reliability of the decision making process, (4) representing in a concise and unambiguous manner standards, codes, and government regulations, (5) providing aids for the use of conventional programs (simulation, finite elements, etc.), (6) getting quicker solutions, and (7) reducing cost.

Clearly, civil engineering students need to take advantage of the methodologies and programming techniques of Knowledge-Based Expert Systems. They need to cross the bridge from building data processing to knowledge processing applications. Recognizing this need, the Department of Civil Engineering at Louisiana State University has been offering a graduate course on expert systems. At first the course was taught as a "special topics course" (CE-7700) with only three graduate students. Subsequently, it was developed as a regular course "CE-7580: Expert Systems in Civil Engineering" and was first taught in the Fall Semester of 1987. Since then, this course has been offered once a year.

This course does not duplicate other (graduate or undergraduate) courses being taught at LSU. While there is some topical overlap with CSC 4444: Artificial Intelligence and Pattern Recognition, IE 4470: Knowledge-Based Systems in Engineering, IE 7470: Intelligent Manufacturing Systems, and EE 4785: Introduction to Expert Systems, this course is specifically designed to serve civil engineering students. A one to one comparison of the above four courses based on the topical outlines of the courses is shown in Table 1. Four important aspects have been compared in this table: (1) computer languages taught, (2) applications covered, (3) emphasis, and (4) prerequisites. Clearly, CE 7580 is different from all the other courses and is designed with civil engineers in mind. This information might be useful to anyone who plans to develop a new course in a civil engineering department.

The enrollment in this course is at about 15 graduate students annually. Academically, students taking this course are about 80 percent from Civil Engineering and 20 percent from Industrial Engineering, Computer Science, Geology, Oceanography, Marine Science, Forestry, Geography, etc.

Although there is a typical prerequisite, the IE-4470 - Knowledge-based Systems in Engineering, this requirement was hardly observed since very few students had taken that course. Some issues related to the prerequisite are discussed in section 8.

The course was developed specifically for graduate students. As the need for expert systems techniques for the practitioner civil engineer becomes more pronounced, it is possible that we will be focusing toward the development of an undergraduate *elective* course.

2 COURSE OBJECTIVES

The objectives of the course are: (1) to introduce the methodologies of artificial intelligence and knowledge-based expert systems, (2) to explore some of the special issues of geotechnical, structural, water resources, environmental and transportation engineering applications, and (3) to provide the students an opportunity to develop a simple knowledge-based expert system. Upon completion of the course, it is expected that the students will be able to: (1) identify a problem requiring heuristic problem solving techniques; (2) create a conceptual knowledge representation of the problem; (3) choose an appropriate expert system tool; (4) formally represent the problem in the selected tool, or language; and (5) test and evaluate the whole knowledge engineering process.

3 COURSE SYLLABUS

Topics	No. of Lectures
Artificial Intelligence and Expert Systems	3
Symbolic Knowledge Representation	9
Rules, Predicate Calculus, Objects, Semantic Nets, Frames	

Feature	CSC 4444: Artificial Intelligence and Pattern Recognition (1)	IE 4470: Knowledge-Based Systems in Engineering (2)	EE 4785: Introduction to Expert Systems (3)	CE 7580: Expert Systems in Civil Engineering (4)
1. Programming Language	LISP	LISP	Prolog	Various Expert System Shells
2. Applications	Natural Language Interfaces	System Design Management, Process Control Manufacturing	Robotics, Computer Vision	Design, Planning, & Interpretation in Structural Hydrological, & Geotechnical Engineering
3. Emphasis	Artificial Intelligence	Industrial Engineering	Electrical Engineering	Civil Engineering
4. Prerequisite	CSC 3102 and IE 3302	IE 4425	EE 3750	IE 4470

TABLE 3-1: Comparative Features of the Four AI Related Courses Being Offered at LSU

Topics	No. of Lectures
Inference/Control Schemes	9
Rule firing, modus ponens, resolution, recognize-act-cycle Forward/backward chaining, depth/breadth first search	
Reasoning with Incomplete or Inexact Knowledge	3
Probabilities, evidences, certainty factors	
Building Expert Systems	3
The architecture of expert systems Steps in constructing an expert system Automated knowledge acquisition, Induction Evaluating an expert system	
Expert System Languages and Tools	6
Language and tools for knowledge engineering Choosing a language or tool Hardware for knowledge engineering Microcomputer tools	
Case Studies of Civil Engineering Expert Systems	6
Water Resources, Structural Design Geotechnical Engineering, Environmental Engineering	
Remote Sensing, Engineering Geology Geographic Information Systems	

4 TEXTS AND REFERENCES

Selecting an appropriate textbook was not easy because most of the available books are targeted towards specific audiences. Lecture materials were compiled from the textbooks and references shown below. Occasionally, journal articles are consulted for specific subjects.

Textbooks:

(1) "Building Expert Systems" by Hayes-Roth, Waterman and Lenat, Addison-Wesley, 1983.
(2) "Expert Systems for Civil Engineers" by Maher, ASCE, 1987.

References:
(1) "Artificial Intelligence" by Winston
(2) "Rule-Based Expert Systems" by Buchanan and Shortliffe
(3) "A Guide to Expert Systems" by Waterman

(4) "Expert Systems in Civil Engineering" by Kostem and Maher, ASCE
(5) "Artificial Intelligence" by Charniak and McDermott
(6) "Artificial Intelligence" by Rich
(7) "Programming Expert Systems in OPS5" by Brownston, Kant, Farrell, Martin
(8) "LISP" by Winston and Horn
(9) "Expert Systems" by Harmon and King
(10) "Artificial Intelligence" by Nillson
(11) "Expert Systems: A Practical Introduction" by Sell
(12) "Introduction to Expert Systems" by Jackson
(13) "A Practical Guide to Design of Expert Systems" by Weiss and Kulikowski
(14) Recent research papers
(15) All the manuals of the expert system tools: XLISP, TURBO PROLOG, VPEXPERT, DECIDING FACTOR, INSIGHT2+, and OPS5.
(16) Many references were given from the following Journals: Computing in Civil Engineering (ASCE), Microcomputers in Civil Engineering (Elsevier), AI Magazine (AAAI), AI Expert (Miller Freeman), PC AI (Knowledge Technology), and other ASCE Journals.

The two textbooks listed above were selected as the primary sources of information. The first provides an excellent treatment of the process of building expert systems. However, it assumes that the reader has had some background in artificial intelligence. I have had to elucidate many of the topics through introductory materials that I compiled from the references shown above. The second textbook provides an excellent treatment of both introductory materials on expert systems and specific applications in civil engineering.

The easiest to read references are: 3, 4, 9, 11, and 12. The materials in references 1, 2, 5, 6, 7, 8, and 10 are more appropriate for those interested in the fundamental aspects of artificial intelligence. I did make extensive use of reference 7 for teaching the fundamentals of production systems by employing the OPS5 language. The OPS5 language was chosen because (1) it is widely available, (2) it is well documented and discussed in depth in reference 7, (3) it offers a flexible forward chaining control strategy, while providing for external control, by system designers, through production rules, and (4) my research prototypes were built in OPS5. Also, extensive use was made of the materials from the Turbo Prolog manual and other chapters of the references regarding logic programming since special emphasis was placed on predicate logic and backward chaining methods.

5 COURSE ORGANIZATION

The course met for two one-and-a-half hour periods per week for lecture. During some of the lectures, demonstrations of the expert systems tools were given in the computer laboratory. The students were required to spend their own time in the computer laboratory for homework and projects.

The course involved homework assignments, a project in the student's area of expertise, an in-class presentation of projects, a midterm and a final exam. The emphasis has been placed on the project.

HOMEWORK ASSIGNMENTS

Selected homework assignments were given from the problems found in the end of the chapters of the following books:

- "The LISP Primer" by B.J. Fladung
- "LISP" by P.H. Winston and B.K. Horn
- "Programming Expert Systems in OPS5" by L. Brownston, R. Farrell, E. Kant, and N. Martin

PROJECTS

The term projects provided actual hands-on experience in building expert systems. They were designed in a manner to require the student to develop from literature review to actual system implementation. The objectives of the various phases of the term project were as follows:

1. Critical reviews of individual papers. To learn to understand and critically review individual papers on advanced technical topics relating to expert systems, students had to read and critically review five to eight papers that were assigned by me or were proposed by them. Written reviews were turned in at the end of the fourth week of class. (10%)

2. Critical reports of inter-relating ideas. To learn to critically inter-relate ideas expressed by several authors on a single subject, students had to write a critical review report comparing and contrasting the ideas expressed in the individual papers read in (1) above. The report was submitted at the end of the sixth week of class. (10%)

3. Bibliography. To learn to develop a bibliography, students had to prepare a recent bibliography in the area treated in

(1) and (2) above. The bibliography was submitted at the end of the seventh week of classes. (5%)

4. State of the art summary. To learn to formulate their opinion about a technical topic, students had to write a paper on the current state of the art in the subject area of parts (1) through (3) above. This statement was not going to reflect their individual opinion regarding the subject area but was going to be based on and documented by reference to the information presented in parts (1) through (3) above. The state of the art summary was due at the end of the seventh week of class. (10%)

5. Research proposal. To learn to plan a research program to advance the state of the art in a technical area, students had to write a research proposal outlining a program involving the design of an expert system as well as theoretical developments that will result in advancing the state of the art in the subject area of (1) through (4) above. The proposal needs to include problem identification, conceptualization and some aspects of formalization of the expert system. The written proposal was due at the end of the ninth week of class. (15%)

6. Implementation and final report. To learn the details of expert system design, students had to:
   a. select one of the available expert system tools,
   b. learn the tool features,
   c. implement their expert system according to (5) above,
   d. test and modify their system, and
   e. write a final report describing the details of their implementation.

The final report emphasized methodological aspects, such as, problem identification, conceptualization, expert system tool description, formalization (knowledge representation, reasoning process, inexact reasoning) implementation, testing and evaluation. (40%)

The results of their work was presented in class during the last three class meetings (30 minutes per student).

EXAMINATIONS

A midterm and final exam were given in the class. Some typical questions from both exams are given in the following.

1. Compare frame and production systems as it concerns their methods for (1) knowledge representation and (2) inference (search and control) strategies.
2. Illustrate by example the distinction between forward and backward chaining and forward and backward reasoning.
3. Explain the difference between "rules about facts" and "rules about rules" (metarules). Provide examples from your major field of study.
4. Briefly discuss conflict resolution strategies employed in popular production system architectures.
5. Illustrate, by example, uncertainty handling methods in expert systems.
6. Discuss the advantages and disadvantages of frames, predicates, and production rules.
7. Discuss features of PC-based expert system tools in the market today.
8. Describe the aspects of problems in your field of study that lend themselves to expert system development. Explain the form of expertise.
9. Discuss aspects of testing and evaluation of expert systems.
10. Discuss the nature of applications of expert systems in your field.
11. Why have expert systems been successful in the market?
12. Discuss architectures or components of ES tools.
13. What are the weaknesses of current expert system tools?
14. What is knowledge (forms, expressions)?
15. What is involved in building expert systems? Explain the steps.
16. What are the fundamental qualities of expert systems?
17. Explain how evidences support hypotheses in PROSPECTOR?
18. What is modus ponens and what is resolution?
19. What are class elements in OPS5 and what are recursive objects in PROLOG?
20. What is depth-first and breadth-first search? What are some alternatives?

6 HARDWARE/SOFTWARE USED

Hardware Tools:

Since no special equipment funding was provided for teaching this class, the Department's IBM AT microcomputers were used. We have been evaluating the situation especially as it concerns the purchase of advanced workstations for teaching and instruction of expert systems. Our recommendations appear in the Future Plans section of the chapter.

Software Tools:

The following software tools were used by the students for implementing their projects: XLISP, TURBO PROLOG, VP EXPERT, DECIDING FACTOR, INSIGHT2+, INTELLIGENCE COMPILER, and OPS5.

XLISP provides a PC language for learning the basics of LISP. Some assignments were given for the practice of basic concepts.

TURBO PROLOG (Borland) provides for logic programming in a declarative form that uses Deductive reasoning in backward chaining form for problem solving.

VP EXPERT (Paperback Software) provides backward and forward chaining, confidence factors, an inductive front end, multiple windows, text, and graphic traces.

DECIDING FACTOR (Channelmark Corporation) provides a very simple decision support system.

INSIGHT2+ offers both forward and backward chaining, confidence factors, help, explanation, and math facilities, and support of external algorithmic programs written in Pascal.

INTELLIGENCE COMPILER employs both forward and backward rules and at the same time provides for use of frames. This last feature makes it one of the best tools in that price range.

OPS5 was used on the VAX/780 computer. It is a production system with forward chaining inference engine. It has two conflict resolution strategies that are not found in other PC tools listed above.

The expert system tools we bought were the kind that we could afford. They were not necessarily the ones we would buy if we had more funds. However, we looked at combining various types of tools instead of centering on one kind, at least at that stage. The selection of the expert system tools turned out to be a good choice because of the experience gained in trying different methods of programming and various tool features.

Two weeks of lectures were spent in explaining some of the main features of each tool. The students had to pick up on the details of each tool for their particular projects. Through my demonstrations of the tool features and the students presentations of term projects, the class was exposed to all the tools.

One of the difficulties encountered was that some of the expert system tools were received after the semester started. As a result there wasn't much time available to develop case studies for instruction. Case studies were developed only in the OPS5 language. However, most of the expert systems tools were not difficult to use and they came with several knowledge bases of differing complexity for learning.

7 DESCRIPTION OF TERM PROJECTS

The following are typical expert systems developed by the students in their term projects.

TAX: An Expert System for Terrain Analysis

The Terrain Analysis Expert (TAX) falls in the area of terrain analysis. Terrain analysis is the systematic study of image patterns relating to the origin, morphologic history and composition of distinct terrain units, called landforms. The goal of a typical consulting session with the Terrain Analysis Expert (TAX) was to infer the landform type of a terrain site appearing on a stereopair of aerial photographs. The approach followed to infer the landform of the site was the landform-pattern element approach. TAX's knowledge was described with models of terrain-related facts and decision rules pertaining to problem solving in terrain analysis. Facts and decision rules with uncertain knowledge sources were identified and methods were developed for their representation. The knowledge base was composed of facts and rules and it constituted the OPS5 program code (Argialas and Narasimhan 1988b). Terrain analysis facts were represented as objects or element classes in the OPS5 language. The problem-solving strategy in terrain analysis was represented through production rules pertaining to the defined concepts.

BUFERSS - ButterFish Expert Remote Sensing System

BUFERSS is a prototype microcomputer based expert system for the interpretation of satellite and research vessel acquired oceanographic environmental data to delineate potential high probability areas for large concentration of fish. It was specifically designed for application to gulf butterfish in the northern gulf of Mexico. The data used are mainly depth, satellite measured sea-surface temperature and pigment, the horizontal gradients of temperature and pigment, and moon phase. The system was developed in the Intelligence Compiler expert system tool. Knowledge representation in BUFERSS employed a combination of frames and rules. The system allows for confidence factors associated with rule premises and conclusions, reasoning under uncertainty, and the use of imprecise or incomplete information. The main output product is a map showing sub-areas of the study area which are likely to have large butterfish concentrations.

SIR: Soil Improvement and Reinforcement Expert

This was a prototype expert system for soil improvement and reinforcement techniques in highway embankment construction. It assists novice geotechnical engineers to select the most suitable method for a highway embankment project. Knowledge representation was based on production rules written in Turbo Prolog. Problem solving was a combination of forward and backward reasoning. The program was structured in three parts: user interface, inference engine, and knowledge base. Explanation facilities have been included in the form of explanation rules to answer user's queries about why the question was asked and how the conclusion was obtained. The program was composed of 51 rules in the knowledge base, 19 rules in the inference engine and the menu driven user interface.

HEC-1 ADVISOR: A Consultation System for HEC-1

HEC-1 ADVISOR was developed to assist the users of HEC-1 Flood Hydrograph Package in formulating the input specific to their problems. The HEC-1 ADVISOR or ADVISOR is a data driven program. The domain of its knowledge base is the HEC-1 input, format, parameters and their units and all the information obtained from the input description manual. The user identifies a problem on the prompt from the ADVISOR. Then based on user response to the questions asked, it queries the user about the essential data to solve the problem. The output file from the ADVISOR can be directly input to the HEC-1 program without any further editing. The ADVISOR was built as a rule-based production system in Turbo Prolog. It employed representation of the decision making alternatives in the form of AND/OR search trees. The knowledge base of the advisor was divided into various subtasks to introduce modularity. These subtasks were formalized as groups of rules and included: 1) Job selection and sub process identification rules, 2) check_ ... rules : check a condition for a subprocess, 3) get_ ... rules (Valid input) : accept data type as needed, 4) .. CARD rules: Rules specific to each card, and 5) General purpose/ Utility rules.

MACERAL: A Prototype System for Maceral Identification

A geology student developed this system for Maceral Identification in Organic Petrology. MACERAL aids in the identification of maceral types and degradation levels using documented classification systems for clastic and carbonate depositional systems. The data are in the form of particle attributes as seen under a transmitted light microscope. It was implemented in VP Expert. It was composed of 17 rules. Some rules were formed by using the internal induction table of VP Expert. Other rules were structured in the editor of the system.

CE-ADVISOR

This system was designed to provide advice to undergraduate students of the civil engineering department in selecting courses, by taking in to account their present status, required prerequisites, etc. It was written in Turbo Prolog and was composed of production rules expressing relationships among courses, prerequisites, etc.

8 COURSE CRITIQUE AND ASSESSMENT

The diversity of student backgrounds can contribute to a larger and more interesting audience which provide a better learning experience for both the teacher and the students. On the other hand, diversity in the students' background might require special attention and can also contribute to a differential teaching effect since some students will be differently prepared in their curricula and will exhibit different analytical, computer, and conceptual skills than others.

Some students had taken the CSC 4444: Artificial Intelligence and Pattern Recognition instead of the assumed prerequisite and some had no prerequisite. We are in the process of evaluating the situation concerning the prerequisite because it does create some problems. For example, if some students have taken the IE-4470: Knowledge Based Systems in Engineering or the CSC 4444: Artificial Intelligence and Pattern Recognition then they are much superior in performance than the rest of the students. It has also been difficult to evaluate the pace of the course since variability in the student population requires adjustments on the part of the instructor. If the course was offered with the strict compliance of the prerequisite, then it would be very hard to find enough dedicated students to take both courses.

In looking back, the factors that made the course successful were: the moral and financial support of the Department, the selection of appropriate textbooks, a fairly easy to use software with good supporting material, an enthusiastic group of students, and the well documented cases of sample applications presented in class.

Although, this course provides the basics of artificial intelligence and expert system design, it does not provide an in-depth treatment of the particular issues that are important for each of the possible applications in civil engineering. Significant enhancement of expert system utilization and integration into the specific disciplines would be expected through interest and assignments given by all the instructors of the civil engineering core courses.

Two research publications were initiated during this course:

> Argialas, D. and Narasimhan, R. 1988a. TAX: A Prototype Expert System for Terrain Analysis. "Journal of Aerospace Engineering," American Society of Civil Engineers, Vol. I, No. 3, July, pp. 74-87.
>
> Argialas, D. and Narasimhan, R. 1988b. A Production System Model for Terrain Analysis Knowledge Representation. "Microcomputers in Civil Engineering," Elsevier Science Pub. Co., Vol. 3, No. I, June, pp. 55-73.

Two more publications are under preparation.

9 FUTURE PLANS

The following discussion describes some of our equipment and software plans of the Department. The Department did acquire software (Deciding Factor, Turbo Prolog, Insight2+, VP Expert) for use on IBM PC/AT but they proved to be inadequate for advanced instruction and research use. Personal computers do offer low-level support for some expert system development applications but fail to offer the computing speed, disk storage, and the operating system support necessary to develop state-of-the-art expert system applications.
Knowledge-Based Expert System developments require such a specialized level of support that, until recently, they were only implemented on LISP machines, like the Symbolics and the TI Explorer, which are custom-architecture computers designed specifically to support Expert System development. They are expensive, and tend to offer minimal support for traditional programming. The Department is looking toward the development of a specialized laboratory for teaching and research in expert systems consisting of workstations and an advanced Expert System Tool. The heart of the system will be an advanced tool like the Knowledge Engineering Environment (KEE) by Intellicorp. Efficient operation of such tools requires a workstation with high resolution, 16" or 19" color graphics monitor, 3-4 MIPS execution rate, 12 to 16 MB main memory, minimum of 110 MB hard disk, the UNIX operating system and a tape drive.

A new undergraduate course in expert systems applications in civil engineering is under development. This course would be designed to address unifying concepts for all civil engineering disciplines as well as provide insight into specific discipline applications. It will also serve as the required prerequisite. Then, it would be expected that major instructional or curricular impacts would result from changes within specific civil engineering courses by introducing assignments and projects that involve this new technology.

Acknowledgements. The author wishes to thank Prof. Roger K. Seals, Chairman of the Department of Civil Engineering at Louisiana State University for his support in the development of this course and for his valuable comments regarding this manuscript. I also thank Dr. Yalcin B. Acar and Dr. James Cruise for their constructive comments. Finally, I am grateful for the helpful comments of Dr. Satish Mohan, Dr. Mary Lou Maher, and the reviewers.

Chapter 4
A Course on Expert Systems in Civil Engineering at Carnegie Mellon University

by Steven J. Fenves, Hon. M. ASCE

1 INTRODUCTION

This paper summarizes a course on Expert Systems in Civil Engineering as it has evolved over five years. First offered as a seven-week minicourse in the Fall of 1983, it has been offered as a full-semester course every year since the Spring of 1985. The course normally attracts 25 to 35 graduate students from Civil Engineering, other engineering departments and Architecture, and a few seniors with advanced standing. The course meets for two 75-minute lectures a week. There is one midterm exam, which counts for 10% of the grade. There is no final exam. All of the lectures are given by the instructor, with occasional guest lecturers presenting expert systems in allied fields.

There are no formal prerequisites. Conventional programming skills in procedural languages are not required. Experience in designing and developing algorithmic programs is helpful. Particularly valuable background and motivation for the course is *frustration* with algorithmic programs, either as a user (frustrated by the "black box" structure of such programs with their built-in, unmodifiable assumptions) or as a developer (frustrated by the users' reluctance or resistance to use such programs). Familiarity with the CMU network hardware and software (operating systems, editors, and text processors) is an absolute requirement.

2 COURSE PHILOSOPHY

This course is part of the core sequence in the Computer-Aided Engineering (CAE) graduate program in Civil Engineering. All the courses in the sequence deal with the application, adaptation and extension of emerging computer-based technologies to practice and research in civil engineering. Specifically, this course deals with the methodology of knowledge-based expert systems, computer programs which use domain-specific knowledge to solve ill-structured problems in that domain, and with the process of compiling, organizing, and formalizing domain knowledge and implementing it in an expert system program called *knowledge engineering*.

Most of the literature on expert systems postulates that expert systems will be developed by a team composed of a *domain expert*, with specialized knowledge about the application domain, and a *knowledge engineer*, with generalized knowledge about expert system building tools, who extracts the expert's knowledge and translates it into the expert system program.

The philosophy of the course is predicated on a different scenario. It is assumed that the background and professional aspirations of the student lies in a specific domain, namely Civil Engineering, and that the student is majoring in CAE because of his interest in learning and applying the latest computer-based technologies to that domain. The course further assumes that through his education, background, concurrent course work, work experience or project experience, each student is at least a *novice* domain expert in some branch of Civil Engineering or a related discipline. Based on these assumptions, the philosophy of the course is that *the best vehicle for introducing a student to knowledge engineering concepts, methods and tools is to have him organize his domain knowledge and translate it into a prototype expert system.* The course thus reflects the CAE faculty's firm belief that expert systems methodology will eventually become just another tool in the kit of computer application-oriented engineers. The course thus differs markedly from courses on Expert Systems and Artificial Intelligence offered by the Computer Science Department which concentrate on domain independent knowledge engineering methods.

Experience with civil engineering students not majoring in CAE and with students from other disciplines indicates that the above philosophy also holds for them. The nature of the domain is not nearly as important as the existence of *some* domain to which knowledge engineering is applied. As a matter of fact, non-CAE students frequently have deeper domain knowledge than the CAE students.

The "deep immersion" philosophy also

applies to the coverage of knowledge-based programming methods or paradigms. One paradigm, rule-based programming, is covered in depth and used by the students in their prototype expert systems. Other paradigms are covered with considerably less depth.

3 COURSE SYLLABUS

The course outline is shown in Table 4-1. After a week's introduction to the concept of expert systems, the lectures are organized into three segments of roughly equal length.

The first segment deals with a specific knowledge-based programming paradigm, *rule-based programming*, and a specific tool in that class, OPS5. The lectures are closely patterned after the organization of the textbook. The objective of this segment is to provide the knowledge-based programming skills that the students will need in the second segment. Based on the experience with the course in previous years, considerable time is spent on structuring of the OPS5 context (working memory) and introducing sample control strategies, such as *sequential, forward chaining and backward chaining*, in this segment. This helps the students get started with programming their knowledge. A mid-term examination is given to let the students test their OPS5 programming skills before proceeding to the next segment.

The second segment deals with the *knowledge engineering process* of developing an expert system application. This segment is primarily intended to provide structure and guidance to the students as they choose, define, design and implement their specific expert system project (see below).

The third segment provides a broad (but necessarily shallow) overview of knowledge-based paradigms and expert systems topics beyond those that the students can incorporate in their projects.

4 TEXTS AND REFERENCES

Over the years, we have experimented with several texts. The majority of the available textbooks are not suited to the philosophy of the course, as they are geared to "general purpose" knowledge engineers, rather than domain-oriented students interested in applying knowledge engineering to *their* domain.

The current text, *Programming Expert Systems in OPS5* by Brownston *et.al.* (Addison-Wesley, 1985) is well suited for the first segment of the course. For the remainder of the course, we provide an extensive bibliography of papers. A very useful reference to the students is a collection of outstanding projects from previous years.

Lecture	Topic
1	Introduction
2	Overview of ES
Rule-based programming	
3	Introduction to OPS5
4	OPS5 Working Memory
5	OPS5 Production Memory
6	OPS5 Process
7	Control in OPS5
8	Control in OPS5
9	Data representations
10	Working memory organization
11	ES global structures
Knowledge engineering	
12	Suitable domains
13	ES development process
14	MIDTERM EXAM
15	Review of project proposals
16	Sources of knowledge
17	Evaluation of constraints
18	User interface, FILL
19	Knowledge base verification
Overview of ES techniques	
20	Knowledge acquisition
21	Generative ES
22	Frame representation
23	Frame representation
24	Logic representation
25	Blackboard architectures
26	Inexact reasoning
27	Geometric reasoning
28	Commercial tools
29	Future of ES

Table 4-1. Course Outline

5 COURSE ASSIGNMENTS

Assignments fall into three categories, corresponding to the three segments of the course. The first three assignments, counting for 20% of the grade, are programming exercises in OPS5 intended to provide the basic skills needed to develop good OPS5 programming practices. The exercises cover the primary facilities of OPS5 in the context of very simple diagnostic applications.

The next six assignments constitute the students' term project involving the conception, design, implementation and documentation of a prototype expert system in the domain of the students' choice. The sequence of assignments roughly parallels the development sequence described in Chapter 12 of [Harmon 85] or Chapter 5 of[Hayes-Roth 83]. The specific domain is up to the student as long as it bears some relation to Civil Engineering (or to the student's discipline, for those outside of the Department) to the point where the instructor and the teaching assistants can understand the application and provide assistance. Students are free to consult faculty and other people for domain expertise, but they are warned that their project will be evaluated as much on the domain expertise demonstrated and incorporated in the system as on the knowledge engineering tools used. Small (<3) group projects are acceptable if prior approval is obtained from the instructor. Use of the project for other purposes (e.g., research project work or exploratory work toward a thesis) is strongly encouraged.

The six assignments and their weights are as follows:

Assignment	Weight
Narrative description	0
Functional specification	5
Summary of domain knowledge	5
Design specification	5
Sample implementation	5
Prototype - term project	30

The first narrative description of the proposed project is not due until the sixth week of the course. However, students are strongly encouraged to begin thinking about the project as early as possible, and use the reference material provided for ideas. Students are encouraged to choose their project topic on their own. A list of typical projects suitable for term projects is provided.

The evaluation of the projects is based on the following criteria:

- *Organization:* selection and implementation of the control structure (goals and subgoals and their execution); structuring of domain rules; organization of working memory;

- *Use of domain knowledge:* extent, quality and validity of domain expertise incorporated; kinds of domain-specific heuristics used;

- *Documentation:* user interface, prompts and explanations; program documentation;

- *Evaluation:* student's self-evaluation of appropriateness, formalism, and prototype scope; and

- *Expandability:* potential for expansion of prototype into a production version.

The third set of two assignments, counting for 20% of the grade, are informal exploratory studies of how the advanced topics covered in the third segment of the course may be applied to the student's project.

6 ENVIRONMENT

OPS5 has been the knowledge-based programming language since the course began. It is an appropriate environment for the kinds of projects expected from the students. The fact that OPS5 basically supports *forward chaining* only is actually an advantage: it permits an explicit treatment of control strategies by implementing a few control - as opposed to domain - rules. Students can implement a variety of control strategies, including *backward chaining* at the level of *goals* and still exploit OPS5 opportunistic inference strategy of OPS5 at the level of individual rules.

In response to request from previous classes, three enhancements to OPS5 have been developed. First, a mathematical function library has been provided for functions more complex than OPS5's *compute* function. Second, the FILL function is provided to facilitate the prompting for input of *all* of the unbound attribute values of an object using only one rule, instead of the students having to write a number of separate rules, each requesting a value for one attribute. FILL requires a separate data structure, built by the student in LISP using some very basic LISP functions, in which is stored an applicable prompt, a type declaration and an allowable range for each attribute. FILL checks the user's input for type and range consistency and re-prompts the user as needed.

The third enhancement, a Decision Table Evaluator (DTE), was developed to facilitate the coding of decision logic tables. A *decision logic table* is a concise way to represent the logic used to arrive at a decision, specifically, the assignment of an attribute value. For example, if the value assigned to an attribute depends on N conditions, the student might possibly have to write as many as 2^N rules to properly assign a value to this attribute. However, if the student uses the DTE function, he only writes one OPS5 rule that calls DTE and builds one LISP-based decision table containing the logic involved in the assignment of a value to an attribute. Thus, the student can again reduce the number of rules in his system, requiring less search by the inference engine and reducing the clutter in the knowledge base. Students are encouraged to use DTE for deterministic or procedural aspects of their knowledge-base, such as table lookups, specification provisions, etc.

The hardware environment has evolved from mainframes (DEC20's) used in the early years to the University's ANDREW distributed environment. Since ANDREW machines are ubiquitous on the campus, access to them is readily available.

7 LEVEL OF STUDENT INVOLVEMENT AND INTEREST

Students enter the course with high anticipation, generated by the publicity about expert systems. After the first few weeks, there is a marked letdown and some thinning out as students realize the amount and kind of work involved. Comments heard while signing course-drop cards are: "I wanted to take *one* computer course at CMU, and thought that this one would require the least work"; or "I did not realize that this was going to be *another* programming course."

Interest and involvement rise considerably as the work shifts to the prototype expert system. Many students present incredibly ambitious proposals, and have to be repeatedly advised to reduce the scope to manageable proportions. A small number of students can't decide on a topic, and have to be guided to focus on a project. This category often includes undergraduates, who are unsure of their domain knowledge; we frequently advise them to work on a tutor for some course they have taken, where they have more expertise than the students entering the course. We are constantly amazed by the range of subjects the students choose. Many of the most successful projects are based on outside experience: a masonry productivity estimator written by a student whose father is a bricklayer; a water quality diagnostician by a student who works as a lifeguard at a swimming pool; a racing bicycle configurer by a student who owns a bike shop; etc.

As their work progresses and knowledge engineering begins in earnest, the students invariably realize that they have considerably overestimated their domain knowledge. We spend a lot of time referring students to faculty, external experts and libraries for sources of domain knowledge.

As an illustration of the breadth of student interest, Table 2 presents the titles of the student projects from the Spring 1988 course.

The discussion of student involvement would be incomplete without mentioning the involvement of the teaching assistants (TAs). The course is run with two TAs. The first two assignments on the term project are read and commented on by the instructor and both TAs. At that point, each student is assigned to one of the TAs, based on commonality of interests, and that TA continues to monitor and assist the student throughout the project. The TAs have also provided and maintained the enhancements to OPS5 described above.

8 COURSE ASSESSMENT

The course is a useful and integral component of the Department's graduate sequence in CAE. Students majoring in CAE acquire exposure and proficiency in a methodology which will increasingly become a significant aspect of computer-aided engineering. The exposure to rule-based programming reinforces the emphasis on good program design techniques and program *modularization* in the other CAE courses. Many good CAE students initially resent "relinquishing" program control to the inference engine (OPS5 interpreter); after a few exercises, they learn to appreciate and exploit the OPS5 control structure, which relieves them of the need to specify in minute detail the sequencing of steps. Non-CAE students, not having been taught otherwise, do not seem to

SHAPEX: An ES for the selection of cross-sectional shape for a structural component

An ES to determine the feasibility of retaining wall rehabilitation strategies

EXMA: A KBES for excavating machine selection and duration estimation

YUST1: Structural steel design processor

Construction project planning aid

RIPET: Remedial investigation preapplication evaluation tool

LOGS: An ES for interpretation of underground profiles from multiple boring logs

Manufacturability evaluator for stamped parts

Tunneling advisor

Die casting diagnostic ES

ES for concrete arch bridge design

Foundation-Expert: An ES for pile foundation design

COMMOD: Concrete mix modification

SILT: Soil investigation and laboratory testing ES

Seismic design consultant

EDAD: Earthquake damage diagnosis

Table 2. Student Projects, Spring 1988

have this hurdle to overcome.

Both groups of students gain a great deal of understanding and appreciation of the meaning of *experience* or *expertise*. They realize that collecting, organizing and formalizing the heuristic knowledge that experts have mentally compiled is a major task; students refer to this aspect of their projects as *knowledge mining*. The students also realize that the available knowledge acquisition facility (i.e., the OPS5 language) is rudimentary, so that the encoding of the acquired knowledge is still a burdensome task.

A measure of the success of the course is that a significant number of student projects from previous semesters have been subsequently expanded through consultation with more mature experts in the specific domains and put to practical use. Many of the prototypes developed in the course have been further refined and enhanced as parts of M.S. and Ph.D. theses.

9 FUTURE PLANS

It is expected that the subject matter of the present course will continue to be presented, but possibly with some significant changes in the mode of instruction. First, as the CAE course sequence becomes further integrated and refined, we may reach the point where knowledge-based methods will become one element of our course on *Elements of CAE Systems in Civil Engineering*, which presently covers elements such as databases, geometric modeling and user interfaces. The development of an expert system would then become a team project in the course on *Design of CAE Systems in Civil Engineering*. Second, there are plans to offer a college-wide course on Expert System development, sponsored by the Engineering Design Research Center, patterned after the present course. Third, if appropriate environments become available, we may offer an expert systems course to non-CAE majors and undergraduates with less emphasis on rule coding and a much "friendlier" knowledge acquisition facility for representing declarative domain knowledge.

Acknowledgements. The evolution and success of the course owes a great deal to the students, and particularly to the teaching assistants, who contributed to the present format of the course. This acknowledgement is only a token of appreciation to M.L. Maher, H.C. Howard, P. Mullarkey, D. Sriram, J.H. Garrett, Jr., N. Baker and G. Turkiyyah for their enthusiastic assistance.

10 REFERENCES

Harmon, P. and King, D., *Expert Systems*. John Wiley & Sons, New York, 1985.

Hayes-Roth, F., D. Waterman and D. Lenat, *Building Expert Systems*. Addison-Wesley, Reading, 1983.

Chapter 5

Teaching a Graduate Course on Expert Systems to Civil Engineers

by Satish Mohan, M. ASCE

1 INTRODUCTION

A graduate course, *CIE 596: Expert Systems in Civil Engineering*, was offered for the first time in the spring semester of 1988 by the Civil Engineering Department of the State University of New York at Buffalo. This course is listed as an elective in the Master of Engineering program in Construction Management and is open to graduate civil engineering students. Graduate students from other disciplines and professionals from the industry can register with the permission of the instructor. This course is planned to be offered every spring semester. The course started with 16 students of which nine were graduate students in civil engineering, two were graduate students in mechanical engineering, one a graduate student in chemical engineering, two engineering faculty, one a computer science major working with the New York State Department of Transportation, and one graduate student in the School of Management. Both of the faculty members, the one computer science major and the one graduate student from the School of Management dropped after four lectures and the first lab assignment. All of them commented that the continued understanding of the course was dependent on the home assignments and lab projects for which they did not have time. The remaining 12 students continued in the course.

There was no maximum limit on class size. Since such a course requires about two to four hours of tutored computing lab where students need individual attention of the instructor and several more hours of office time for meeting students regarding their projects, it is suggested that the class size should be capped at 20 unless a teaching assistant with adequate experience in the expert system software(s) to be used in the course is made available.

The course had no prerequisites because it was expected that the graduate students interested in learning expert system techniques would have adequate background in the use of computers. However, more than half of the students had not used computers. The first lab was therefore spent on basics of PCs, DOS, and Lotus 1-2-3™. All of the students gained the necessary working knowledge very quickly and seemed comfortable with their computers in the second week. As the course progressed, it was noticed that the students who had adequate experience with computers spent less time in the computing lab and built more capable expert systems in their class projects. It is, therefore, suggested that a first course in computer science should be a prerequisite for a graduate course on expert systems.

2 COURSE OBJECTIVES

Expert systems are one of the major applications of Artificial Intelligence -- an emerging technology with a high potential of solving business world problems more efficiently. In the last few years, many expert system applications have been developed and are in routine use. In the field of civil engineering, however, there are very few operational expert systems. Much of civil engineering is experience based and can use the expert systems techniques for advancement. Also, most small and medium civil engineering businesses have not so far been able to reap the benefits of computer technology because most of them do not have knowledge of computers and cannot afford to employ computer scientists. Expert systems provide natural language user-interfaces and will thus be very attractive to engineers and managers who have no computer science background. On the other end, expert systems can be integrated with numeric and other algorithmic programs, databases, and graphic packages, thus providing the benefits of exact design/optimization methods. Civil engineers now have a never-before opportunity of high technology advantages in expert systems and not-to-miss opportunity of the revival of their leadership tradition which they have lost in the past two decades. With this express need in mind, this course on expert systems was designed with the sole objective of preparing graduate students in civil engineering to build prototype expert systems in the domain of their interest and expertise. Using commercially available expert system shells, they can incorporate the expert systems technology in their work as civil engineers.

3 COURSE SYLLABUS

The course contents were distributed into 16 weeks as in Table 1.

TABLE 1: CIE 596 - Expert Systems in Civil Engineering Course Syllabus

Week of	Topics (Suggested Reading)
Jan. 25	Course Introduction Overview of Artificial Intelligence - Historical Development - Basic AI Concepts & Techniques (Chapter 1 - Text)
Feb. 1	Expert Systems - Characteristics - Expert Systems versus Algorithmic Programming - Operational Expert System Examples System Architecture (Mohan's Paper)
Feb. 8	M.1. Knowledge System, M.1 Videos
Feb. 15	Knowledge Representation Schemes - Rules (Chapter 4 - Text)
Feb. 22	Knowledge Representation Schemes - Semantic Nets - Frames Uncertainty & Fuzzy Knowledge
Feb. 29	Logic & Theorem Proving - Predicate Calculus - Boolean Logic Inference Nets
Mar. 7	Control Strategies - Forward Chaining - Backward Chaining - Examples (Chapter 5 - Text)
Mar. 14	Search Methods - Depth First - Breadth First - Search Efficiency
Mar. 21	Knowledge Elicitation - Interview - Protocol Analyses - Induction - Repertory Grid Technique - Mycin's Knowledge Base
Mar. 28	Spring Recess
Apr. 4	An Operational Expert System Example
Apr. 11	Expert System Building Tools & Languages - Symbolic Programming Languages (LISP, PROLOG) - Knowledge Engineering Environments (KEE, S.1, ART, KNOWLEDGECRAFT) - Expert System shells (M.1, PC+, INSIGHT2+, KES) (Chapter 7, 8 - Text)
Apr. 18	Expert System Building Process (Chapter 11, 12 - Text)
Apr. 25	Survey of Operational Expert Systems (Chapter 9, 10 - Text)
May 2	Project Presentations
May 9	Future of AI & Expert Systems

The first week's two lectures introduced the course and gave an overview of how AI has evolved and what are its current applications in engineering. An expert system shell DECIDING FACTOR™ was also introduced in the first week. The second week included a discussion on expert systems architecture and other characteristics as compared to conventional programs. Examples of three rule-based expert systems (MYCIN, DRILLING ADVISOR, and PROJECT PLANNER) were presented in this week to illustrate the various expert system characteristics, earlier discussed. The course then proceeded on to teach the various knowledge representation paradigms and the treatment of uncertain and fuzzy information. In the first five weeks, the students had turned in their first prototype expert system using the DECIDING

FACTOR™ shell and had started on the M.1™ expert system shell. The various distinctive features of expert systems were now transparent to them.

The next three weeks covered predicate calculus, control strategies and search methods. During this phase, small illustrative examples were found very useful in explaining the precise differences between the various search methods and control strategies. Towards the end of these weeks students were in the middle of their second project using M.1™ shell and were acquainted with almost all A to Z steps of an expert system building process.

The next six weeks were devoted to enriching, formalizing, and discussing advanced features of expert systems, as used in some of the operational systems. This period included two lectures on knowledge acquisition and two lectures on tools and languages. One remaining week of the semester was then used in the presentation of students' projects and the course concluded with a lecture: *Future of AI & Expert Systems*.

4 TEXTS AND REFERENCES

The following three books were used in this course.
1. Harmon P., and King D., *Artificial Intelligence in Business: Expert Systems*, John Wiley & Sons, 1985.
2. Winston P., *Artificial Intelligence - Second Edition*, Addison-Wesley, 1984.
3. Waterman D.A., *A Guide to Expert Systems*, Addison-Wesley, 1985.

Harmon and King's book was used as the course text and the other two as general references. The prescribed text included only a small part of the material covered in the course. The above three books combined also were found insufficient in coverage of topics and were having different foci than the course syllabus. Thirteen technical articles and two papers authored by the instructor were handed out or assigned for library reading during the course of the semester.

5 COURSE ORGANIZATION

The course delivery was organized into two 50-minute lectures and one two-hour tutored computer laboratory each week. Students were expected to devote at least 4-6 additional hours per week in the lab, which they did. This delivery plan worked well. One two-hour lab in the middle of the semester was used for students' presentation of their final proposed systems and one lab towards the end was used for presentation of their completed prototypes. Both of these presentations were voluntary and about half of the students participated each time. These presentations were very interesting and proved rewarding to most in that they triggered discussions between students on several knowledge engineering issues.

The course requirements included seven individual homework assignments, two minor projects and one major term project. No in-class exams were required. The homework assignments were based on class lectures and suggested readings and included problems on rule-formulation, truth tables, consistency checks, problem solving, control strategies, search techniques, certainty factors and knowledge representation schemes. One last homework was an essay on knowledge acquisition issues.

Since the major objective of this course was to prepare students for building expert system prototypes using shells, this course required development of three expert systems during the course starting from a very small system using DECIDING FACTOR™. The next two systems were developed using M.1™ shell, the first one was required to have 20-50 rules and the second one to have close to 100 rules. The lab schedule was planned as in Table 2, below:

TABLE 2: Computer Lab Schedule

Week	Assignment
1,2	DECIDING FACTOR™ Software Project I
3,4,5,6	M.1™ Software - Review of Example Systems (WINE, PHOTO ADVISOR, SACON) - Modify one Example System
7,8,9	M.1™ Software Project II
10,11,12,13	M.1™ Software Project III
14	Project Presentations

The course grading was based on homework and projects and was weighted as follows.

Homework Assignments	20%
Project I (DECIDING FACTOR™)	20%
Project II (M.1™)	20%
Project III (M.1™)	40%

6 COMPUTING ENVIRONMENT

In the fall of 1987 when this graduate course on expert systems in civil engineering was planned, the department's teaching lab had ten IBM PCs with 256k of memory and with no hard disks on them. This was considered very deficient. Also, the department had no expert system software. A budget of $10,000 was proposed and approved for equipping the lab with the most essential items for this course as follows.

(i) Upgrading four PCs to 512k and 20 MB hard disks
(ii) M.1™ Expert System Software - Instructor's package (10 copies)
(iii) DECIDING FACTOR™ Expert System Software (2 copies)
(iv) PERSONAL CONSULTANT PLUS Expert System Software (2 copies)

Because of budget limitations, only four PCs were upgraded. It was expected that some of the students would form groups during the tutored lab periods or work on existing PCs, but it was not feasible. Therefore, the lab had to be moved to a computer science department lab facility which had twenty-five 512k microcomputers networked to two IBM PC/ATs. This worked out very well.

Only two of the three expert system software packages acquired for this course could be utilized, DECIDING FACTOR™ in the first two weeks and M.1™ in the remainder of the semester. It is suggested that in a graduate course such as this where the major emphasis is on building expert system applications, a maximum of two types of software should be used. One of them should require the knowledge represented in rules and the other in frames or objects. Both of the software packages should be capable of integrating with programs external to the expert system software and the class projects should require such linkages.

7 STUDENT INVOLVEMENT AND INTEREST

Most of the students were impressed with the expert systems technology and immediately started seeing its potential in engineering applications. They learned to use the software very quickly and devoted more time on their projects than expected. They were given the choice of selecting topics for their projects, the only warning given was to keep the domain narrow so as to be manageable and so that they could visually see the chaining of various rules during consultation sessions. A selection of their term project topics is listed in Table 3.

TABLE 3: Selected Term Project Topics

1. An Expert System for Determining the Value of Impact Factor for the Design of Highway Bridges
2. An Expert System for Predicting the Failure of a Machine Part
3. An Expert System to Aid in Deciding the Window Orientation and Window Material in Hot Areas of USA
4. PAVEMENT MANAGEMENT ADVISOR-I: A Knowledge System for Recommending the Best Rehabilitation Treatment and its Associated Cost.
5. An Expert System for Chemical Process Control
6. Surface Parking Design Advisor
7. Animal Identifier
8. An Expert System for Evaluating the Feasibility of a Project based on Utility Theory

Students were given the choice of presenting their expert systems in the class during the final two-hour lab and half of them chose to do it. This session was very interesting. The questions asked by the students during the presentations indicated the extent of learning expert systems techniques and the level of interest this course had generated. The results were very satisfying to the course instructor who felt that the course had fulfilled the set objectives.

Five of the 12 graduate students later applied expert systems techniques in their graduate research. One student, D. Baker, used his term project in his M.S. thesis and published a paper entitled, "HIBIC: An Expert System for Highway Bridge Dynamics", in the ASCE Journal of Computing; two students are planning to use expert systems in their Ph.D. dissertation; one student's system has been used in a research proposal and he is building another one for his M.E. project; and another student is linking his expert system to a graphics package.

8 CONCLUDING REMARKS

(i) Based on the experience of teaching the above described course it can be concluded that a one semester course with 20 one-hour lectures on the theory and building process of expert systems, six guest lectures on civil engineering applications and a 2-hour tutored lab per week is considered adequate for a graduate course for preparing students to develop expert system applications and to enable the motivated students to advance through self-learning. Some of the serious students used expert systems techniques in their master's thesis after one such course and did a very good job. For those students, however, who want to use AI techniques in their Ph.D. dissertation, the following sequence of three courses is suggested:

1. One course in an AI programming languages: C, LISP, or PROLOG

2. An introductory course in AI

3. A course on Expert Systems

The first two courses are usually offered by the Computer Science Departments. If a civil engineering department decides to include such courses within their academic program, it is possible to organize the contents of the above three courses into two.

(ii) Students who had some computing experience performed consistently better in home assignments and class projects. A basic course in computers should, therefore, be a prerequisite for a course on expert systems.

(iii) Students learn the techniques of building an expert system using a shell very early in the course. The real problem they then face is the knowledge elicitation from human experts. In spite of constant suggestions to capture human expertise in their final term projects, only two out of 12 graduate students used human experts to fill a part of their knowledge base. Others used books, journals and magazines for domain knowledge. It is suggested that more time should be given to knowledge acquisition methods early in the course. Transcripts and/or videos of a few real efforts in knowledge acquisition would be helpful.

(iv) Most of the affordable commercially available expert system shells are rule-based. Some of the systems developed in the class projects had 100 rules with a seemingly narrow domain and still looked very limited during consultation. Useful looking and interesting systems were those that integrated database management programs and conventional programs. It is suggested that the final term projects should require integration with programs external to the expert system shell, such as spreadsheets, databases, graphic packages or numeric programs.

Chapter 6

A Course on Expert Systems in Civil Engineering at Texas A&M University

by John M. Niedzwecki, M. ASCE

1 INTRODUCTION

With the rapid developments in computer technology, some of the Civil Engineering faculty and students at Texas A&M University became quite interested about how these new technologies would impact the practice of civil engineering in the future. In response to this curiosity, an elective course dealing with expert systems in Civil Engineering was organized and first offered in the spring semester of 1986. Two faculty jointly developed and coordinated this course. The course was designed around four topical areas of interest, specifically: 1. artificial intelligence, 2. computer hardware, 3. computer languages and, 4. applications in civil engineering. In planning the course, it was decided that a two-credit seminar/laboratory format for the course would be the best option and, that because of the breadth of the topical areas, an inter-disciplinary team of university faculty, representatives from the computer industry and, practicing engineers already using these technologies would be needed in order to provide meaningful coverage of the topical areas. Of particular interest were those technologies which would eventually impact personal computers (PC's) and inexpensive workstations, both of which are an integral part of most large or small civil engineering firms. More practical issues included the purchase of software, the upgrading of PC's for student use and travel funds for invited lecturers from academia and industry. All of the necessary funds were provided by the Department of Civil Engineering.

During this same timeframe, the Computer Science Department was experiencing a tremendous growth and could not entertain an additional course of the type being proposed. However, several faculty from the Computer Science Department agreed to give a single lecture on a focused topic and were quite interested in talking with the course participants. The representatives from the computer industry which were contacted were also very supportive. They sent literature and went to great lengths to provide on-campus demonstrations of their latest hardware and software. Invited lecturers from academia were also quite supportive, as was the case with practicing engineers. Clearly, the major problem was to focus the lectures and provide a balanced coverage of the topical areas. Each of the lecturers agreed to discussions with the faculty in charge in order to focus their lectures for the intended audience. The lectures were open to all the Civil Engineering faculty and students but, the size of the laboratory sections were kept small.

The overall faculty response, to an elective course entitled **Expert Systems in Civil Engineering**, ranged from excitement, shared by the course developers, to a lack of interest. As it is today, many people view the whole Expert System exercise as fancy programming with little intellectual content. On the other hand, the students were quite excited and eager to participate, especially in the laboratory portion of the course. Knowing beforehand the range of views, the course developers made a concentrated effort to address these issues and advertised each of the lectures and its content prior to its presentation. It might also be noted that none of the lectures lacked for spirited discussions, which often lasted as long as the lecture. Plans to offer a three-credit version of this course on a regular basis are underway.

2 COURSE PHILOSOPHY

The course was designed with two major objectives in mind. First, we wanted to introduce our senior level undergraduate and graduate students to new and emerging computer technologies. Secondly, we wanted to focus their problem solving activities on the application of knowledge-based technology to Civil Engineering. This course was intended to be an introductory course which would provide students with a solid view of the material while providing "hands-on" experience with an expert system shell. Of course an underlying objective was to stimulate faculty and student interest enough to begin considering research in this area.

3 COURSE SYLLABUS

The course syllabus presented in Table 1. was developed at Texas A&M University for a course entitled: **Expert Systems in Civil Engineering**, first offered in the 1986 spring semester, as a two-credit elective course. There were fourteen lectures and an equal number of laboratory sessions. Additional technical material was presented in the laboratory which was needed by the students in order to develop and complete their term projects.

The lecture topics, presented in Table 1., can be divided into three categories, depending upon whether they were given by Civil Engineering faculty, Computer Science faculty or industry representatives. If viewed in this way, there were seven lectures given by Civil Engineering faculty, three given by Computer Science faculty and four by industry representatives. Of the seven engineering faculty lectures, two were given by Civil Engineering faculty from two other universities. Thus, a total of 9 lectures (64%) were given by speakers from outside of the Civil Engineering Department at Texas A&M University. In addition, each of the students attended fourteen laboratory sessions conducted by the faculty in charge of the course.

No textbook was selected for use in this course. Each of the lecturers was asked to provide any reference material they deemed appropriate. The instructors did make copies of the Radian Users manual available to the students.

4 COURSE ORGANIZATION

The course was developed for a Tuesday/Thursday meeting schedule. The lectures were scheduled for Tuesday afternoons and the laboratory portion was scheduled on Thursday. The students taking the course for credit were expected to attend the lectures and one of the separate laboratory sessions focusing on the development of an expert system in structures or water resources. Each of the students were required to develop a simple

Lecture Topic	# Lectures
Organizational Meeting	1
Artificial Intelligence and Expert Systems	1
Engineering Applications of Expert Systems	1
An Expert System Shell -- Rulemaster	1
Using Rulemaster -- An Industry Perspective	1
Expert Systems -- A Computer Science perspective	1
Comparison of Expert System Shells	1
Comparison of Computer Languages	1
Expert Systems for Hazard Waste Management	1
Knowledge-Based Laboratory at TAMU	1
An Overview of Fuzzy Set Theory and Its Applications	1
Hardware for Expert Systems Applications	1
Expert Systems in Structural Engineering	1
Applications in Civil Engineering	1
	14

Table 6-1. Expert Systems in Civil Engineering Course Syllabus

expert system in the area of structures or water resources and to make a brief presentation on their project at the end of the course.

The course enrollment was limited to eight students, four each in structural engineering and water resources. Two of these eight students, one in each area, were undergraduates. These speciality areas were the primary interest areas of the two faculty in charge of the course. The students interested in structural engineering were directed by Dr. Niedzwecki and were encouraged to work on independent projects. The students interested in water resources were directed by Dr. Strzepek and they elected to collaborate on a single project. The two groups did not meet at the same time. For homework, the students were given technical papers and other technical material to read in addition to the manual on the expert system shell they were to use in their projects. The students' final grade was based upon their project and the presentation of their work.

5 HARDWARE/SOFTWARE USED

The computer equipment, available for the students enrolled in the course included two IBM PC's which were upgraded with 20MB hard disks, two AT&T 6300 series PC's and, one AT&T 3B2 system. The software for these systems was upgraded to make them nearly equivalent in terms of computer languages, spread sheets and data bases.

Rulemaster™, an expert system shell, was selected for use in the course. It was selected because of its suitability for engineering problems and the willingness of the company to work with us. The Rulemaster software consists of an automatic rule generator and a structured rule language called Radial. Further, this software tool provided the user with the option to convert the entire Radial code into an equivalent C language code. This provided code transportability and ease in combining numeric computation written in other languages. Actually, it is possible to call for numeric computations in Radial. The software was capable of forward and backward chaining control strategies and included some limited fuzzy operators for dealing with uncertainties.

6 LEVEL OF STUDENT INVOLVEMENT AND INTEREST

The students were directed to consider developing projects in one of two broad topical areas. More specifically, the use of expert system technology to build design advisors, for routine and innovative design problems, and tutors, for providing shallow and deep levels of knowledge for focused technical topics. These types of expert systems can be useful to both universities and Civil Engineering firms for capturing and presenting focused engineering knowledge, and improving the technology transfer between the university community and practicing engineers. A factor that distinguishes engineering applications is the need to integrate numeric computations of varying complexity with the processing of symbolically represented engineering knowledge. In the structures group the students were presented with general guidelines for developing an acceptable topic but, were required to select their own problem. Two examples of expert systems developed by students follow.

6.1 TUTOR APPLICATIONS

One student decided to build a simple expert system tutor which used the singularity function method for solving beam deflection problems. Of course the development of a comprehensive tutor on this subject was beyond the scope of the course however, what we wanted to build was a simple model, as a proof of concept, which could be used to probe the subtleties associated with building a robust tutor. Thus, it was decided to focus upon the deflection of a simply supported beam subject to various external loadings which could be selected by the user. The processes of problem definition and implementation were quite instructive since, they forced us to recast our solution strategies in a very structured fashion. Issues concerning the sequencing of input, interactive capabilities, internal symbolic processing of the equations, error detection and tutor functions of advice and correction, all came to light. A result, which could have been anticipated, was that when the completed expert system was run as a demonstration for visitors, the expert system technology was basically transparent to the visitors.

6.2 DESIGN ADVISOR APPLICATIONS

Another example developed by a student dealt with an expert system for routine wave force design problems. The expert system was designed to guide the user through the process of selecting the appropriate wave force model for slender pile or monotower structures. An interesting implementation problem which developed was the conflict between shallow and deep knowledge requirements for design. As the student worked with the implementation process the student decided that more knowledge was available which could be crucial to the designer, so the expert system kept growing in scope. This problem illustrated how our thinking changes as we condense engineering design knowledge and, how we naturally develop these systems from shallow to deep knowledge-based systems.

7 COURSE CRITIQUE

By the end of the semester, many questions about computer hardware and software had been answered and some new problems for further research had been identified. The attendance at the lectures typically varied between 25 and 35. The visitors included student and faculty in other Engineering departments and many of these visitors attended most of the lectures. Based upon feedback from both the students and faculty, there is a general consensus that the course was very informative and useful. I think everyone benifited from the lecture portion of the course however, there were some hardware and software related problems which caused some delays. The course laboratory is one area where significant improvements are being made. The goal is to develop an advanced classroom environment capable of CRT projection with six to eight workstations networked to a file server and connected to the campus EtherNet. Several additional expert system shells and supporting software have been purchased. More faculty are becoming involved in expert systems related research and are contributing to this effort.

The student projects required a significant amount of time but, neither the faculty nor the students seemed to mind. It is clear that we should have started the course with programmed examples which the students could modify and learn prior to beginning their projects. A text written for engineers would have been invaluable but, we could not find one at the time. The term projects also forced the students to talk with the software vendors staff and this seemed to motivate some of them to learn the fundamentals of the computer language C. LISP and LISP based shells were ruled out because of cost. In the last few years this has changed and we now have this capability. A major issue facing faculty teaching this type of course is the selection of the software to use in the course, with the range of computer operating systems, computer languages, and shells presently available this is not a trivial problem. For this course we chose an expert system software tool, Rulemaster^{TM}, which was strongly connected to the C computer language. Also its rule-language "Radial" was easy to learn for developing rules from engineering knowledge. Thus, our not imposing any formal course prerequisites and assuming only a mastery of FORTRAN was reasonably adequate.

Both the faculty involved have pursued the subject matter much further. When the course was offered both faculty had little experience and used the course to expand their knowledge of the subject. Dr. Niedzwecki was later supported by the Dean of the College of Engineering to work with faculty at the Massachusetts Institute of Technology who were participating in MIT's Project ATHENA. The collaborative research venture named Project PENSAR focused upon the development of educational software for engineering science and design. Software from both of these projects is now available for use at Texas A&M University. Thesis research and demo-packages developed under Project PENSAR have been reported at technical conferences and journal articles are currently being reviewed. Research work continues on intelligent tutoring systems and design advisors for routine and innovative engineering design applications. Much of this work will be integrated into the laboratory portion of this course on expert systems in Civil Engineering.

8 FUTURE PLANS

A three-credit version of this course is under development which will be directed primarily at our junior and senior civil engineering undergraduate students. It will be open to our graduate students but, it is felt by the author that undergraduate engineers should take this type of course prior to their capstone senior design course. The other undergraduate engineering programs on campus will be notified of the course and, hopefully, some students from other engineering disciplines will participate. The only major change to the course content, will be in the expansion of the laboratory protion, which will now include a more formalized lecture and previously developed examples. Course notes to be purchased by the students enrolled in the course are currently being prepared. Homework will be assigned which requires modification to these existing small expert systems. With the recent purchase of high end personal computers, several workstations and software, the enrollment will be allowed to grow to twenty students. As before, all formal lectures will be open to interested colleagues and students. A more regular schedule for offering the course is planned.

Acknowledgements: The author gratefully acknowledges the support of the Department of Civil Engineering especially, Dr. Donald McDonald who actively supported this initiative. The author would also like to thank Dean Herbert H. Richardson of the College of Engineering who sponsored Project PENSAR. The support of Radian Corporation and Texas Instruments is also gratefully acknowledged. The excellent lectures by Professor James T.P. Yao, now Head of the Department of Civil Engineering, Professor Mary L. Maher of Carnegie-Mellon University, Mr. K. Ferda of the International Institute for Applied Systems Analysis of Vienna, Austria and, Drs. L. Carlisle, P. Freal and D. Friesen, of the Computer Science Department at Texas A&M University, were most appreciated. The author would like to thank the students for their enthusiasm and hard work, especially Mr. Gary McLoskey and Mrs. Katherine Kraft. Finally, the author would like to acknowledge Dr. Kenneth Strzepek, now at the University of Colorado in Boulder, for his cooperation and unbounded enthusiasm during the co-development of this course.

Chapter 7
Teaching Expert Systems to Civil Engineers at the University of Washington: An Opportunity for Assessment

by Richard N. Palmer, M. ASCE

1 INTRODUCTION

Serious concern has been voiced during the past decade over the professional role that civil engineers will assume in the society of the future. Having enjoyed an expanding discipline in the sixties and seventies, civil engineers have endured a contracting employment market in the eighties. Enrollments in civil engineering departments at most universities have decreased and efforts are required to maintain the high quality of students that the profession has enjoyed in the past. One reason for the decrease in enrollment and quality of students being attracted to civil engineering is the perception that civil engineering is not a "high-technology" field. In the minds of most engineering students, high technology is limited to areas such as computer science and electrical engineering where computer technology and computer programming play a major role. Most students do not perceive the significant role played by computing and computers in civil engineering.

At the University of Washington efforts have been made by the faculty to correct this perception. These efforts center on the development of departmental "course-ware" (computer software and exercises written specifically to demonstrate engineering principles and as illustrative homework assignments) for the junior curriculum. By requiring that computing be a fundamental component of all junior level classes, it is anticipated that students will adopt a computer oriented problem solving approach to their senior and graduate level courses.

It is in conjunction with this effort that a course in "EXPERT SYSTEMS FOR CIVIL ENGINEERS" was first conceived. The original purpose of such a course was to demonstrate to civil engineers new computer technology that seemed particularly appropriate to civil engineering applications. What made a course in expert systems especially compelling was the fact that it provided a means by which civil engineering students could be encouraged to consider how subjective and non-quantitative information is used in engineering design and decision making. This is an area in which students, especially undergraduate students receive little, if any, formal training.

2 DEVELOPMENT OF EXPERT SYSTEMS EXPERTISE

Several members of the civil engineering faculty first demonstrated interest in expert systems by participating in the purchase of a site license of M.1. Although this expert system shell was not formally used as a teaching tool, the software provided the faculty the opportunity to explore the use of expert systems and to evaluate its potential for application to civil engineering problems. (M.1 was found lacking by a number of the faculty and it was generally abandoned as a research or teaching tool.)

As is often the case, the skills necessary to teach a course in expert systems were developed through funded research efforts prior to the development of a formal university course. Two faculty members (Drs. Brian Mar and Richard Palmer) had participated in a multi-year research contract with the Boeing Aerospace Company of Seattle to evaluate computer hardware and software that might prove of value in increasing the productivity of its system engineers. As a component of that effort, twelve expert systems software packages were identified and later seven were formally evaluated. In addition, a short-course was developed for introducing system engineering managers to the potential of expert systems for performing system engineering tasks. This experience illustrates the symbiotic relationship that exists between research and teaching and the educational value of such funded research. Without the experience gained from the funded research, any university course on expert systems would have been taught without practical experience of the difficulties related to acquiring knowledge from experts and transforming this knowledge into functioning expert systems.

Experience and expertise gained through the research effort with Boeing also provided the faculty with the expertise to select software appropriate for teaching an introductory course in expert systems in the civil engineering department and an understanding of the potential for expert systems in engineering environments. Concurrently with the Boeing effort, other faculty developed research interests in expert systems that provided expertise within the civil engineering department in a variety of application areas including engineering automation, drought management, wind engineering, and engineering mechanics. A recurring theme in many of these projects is the integration of expert systems with other, more conventional programming procedures, such as simulation, optimization, graphics, and database management (Johnston and Palmer 1988, Mar and Palmer 1988, Miller 1988, Palmer 1987, Palmer and Holmes 1988, Palmer and Mar 1988a, Palmer and Mar 1988b, Palmer and Tull 1987, Reed 1988a, Reed 1988b).

3 COURSE OFFERING

A course in expert systems in the Department of Civil Engineering was initially offered for two reasons. The primary reason for its offering was that expert systems offered an attractive method for incorporating uncertainty, subjective information, and rules of thumb into engineering problems. Secondly, expert systems represented a new computer technology that is likely to play a major role in decision making, analysis, and engineering design in the future.

The faculty believed that a course designed around the needs of civil engineers with examples in their application areas would provide the best vehicle for illustrating the concepts of expert systems. Only through the presentation of relevant examples would the potential of expert systems be fully recognized. In addition, only through the use of class projects would the students become intimately familiar with the advantages and disadvantages of expert systems.

Another practical reason for offering a course in expert systems for civil engineers was the large demand for courses taught on artificial intelligence (AI) and expert systems in other departments. Because of this demand, students in civil engineering found great difficulty in gaining registration to these classes through other departments. (Typically, students within a department are given higher priority for a class taught within their department.)

A course in expert systems was first offered in the Department of Civil Engineering in the Winter of 1986. As with many first offerings taught on the initiative of the faculty, this course was viewed as a teaching overload and taught in addition to the normal departmental requirements of the instructors. The target audience for the course was senior undergraduates and graduate students in all areas of civil engineering interested in the application of expert systems. The prerequisite for the class was a senior or graduate level standing. The class size was limited to 25 students. It was assumed that all students were familiar with personal computing, although knowledge of expert systems shells or AI programming languages was not assumed. Guest lecturers from industry and other faculty within the department are called upon to demonstrate recent research accomplishments.

4 COURSE OBJECTIVES AND PHILOSOPHY

The expert systems course is taught from the viewpoint that the student should learn about expert systems relative to the field of civil engineering. Because simulation and optimization have an important role in this discipline, expert systems are contrasted with these more conventional forms of programming. It is not suggested that expert systems will replace conventional programming, but rather that expert systems offer strengths in problem solving that compliment those offered by traditional programming techniques.

A primary tenant of the course is that students learn through repetitive practice and application of the techniques described in class. Because of this, assignments are a regular feature of the course and examples of programs are discussed during class. Emphasis is placed on the development of simple expert systems and little time is spent on the important topic of knowledge acquisition. This does not imply that the students are unaware of its importance but rather that they do not have the professional experience to appreciate its nuances. Emphasis is also placed on developing systems that interface smoothly with other programs.

Another important reason for offering a course in expert systems is to allow students the opportunity to study the way in which decisions are made in an engineering environment, particularly those related to design, where subjective information is important. In most other courses, students are given techniques for problem solving that rely on objective information and analysis. It is important to train students in techniques that allow more subjective information to be included in the problem solving process and to recognize where such techniques can be used. Students are reminded of the significance of rules of thumb, safety factors, design paradigms, and experience in civil engineering. It is argued that expert systems provide an ideal mechanism for incorporating such subjective information into engineering analysis.

5 COURSE ORGANIZATION

The expert systems course meets for three hourly lectures per week. No laboratory sessions are scheduled, although computer demonstrations are presented with the lectures with the aid of screen projection equipment. The course requirements include two modeling projects, a mid-term and final exam, and weekly computer assignments. The projects are due at mid-term and at the end of the quarter and consist of developing two expert systems for different application areas of civil engineering. Whereas the examinations in the course are designed to emphasize basic concepts in expert systems, the projects are designed to provide the students an opportunity for creative programming in a topic of their choosing.

The course syllabus is presented in Table 1. As the syllabus indicates, the course uses a variety of textbooks and readings. The course requires significantly more reading than is typical of other courses in the department. This need arises from the lack of formal training civil engineering students have in topics in AI and the need to provide the students with a good understanding of the important principles in the field. Emphasis is placed upon reviewing applications of expert systems to civil engineering practice. The ability to accomplish this has been greatly enhanced by the number of ASCE publications that have appeared in recent years devoted to this topic.

The primary text is the class is "Expert Systems: Artificial Intelligence in Business", by P. Harmon and D. King. Although the book is somewhat dated and does not contain civil engineering examples, it does provide a good, non-technical introduction to the topic. "Expert Systems for Civil Engineers: Technology and Application", edited by Mary Lou Maher is the second primary text used in the course. The book illustrates the concepts in the Harmon and King text in a civil engineering setting. "PROLOG, Programming for Artificial Intelligence", by I. Bratko is used as an introductory text for the PROLOG language. This is an excellent text and, unfortunately, there is insufficient time to cover all of its topics and examples. To introduce specific topics related to expert systems and the INSIGHT2+ software, class notes have been written in the form of "A Short Course on Expert Systems for Civil Engineers", by Richard N. Palmer.

In addition to these texts, students are encouraged to review articles from other books and journals to aid them in the selection of interesting design projects. Included in this group is the Journal of Computing in Civil Engineering, IEEE EXPERT, AI EXPERT, and texts edited by Adeli (1988) and Kostem and Maher (1986).

The first week of the quarter is devoted to an introduction to expert systems. The primary objective of these lectures is to contrast expert systems to conventional engineering programming. This is best accomplished with a brief introduction to the PROLOG language. Whatever may be the shortcomings of PROLOG relative to LISP in the development of large, complex systems (and some would argue there are few), PROLOG offers the beginning student the most direct and accessible understanding of logic programming possible. Students are

Week	Topic	Reading Assignments
1	Introduction to Expert Systems Contrasts to Conventional Programming Logic Programming in PROLOG Project Description	(A) pp. 1-14 (B) pp. 1-33 (C) pp. 1-26
2	Recursion in PROLOG Simple Expert Systems in PROLOG Introduction to Insight2+ and Programming Languages Students Turn-in Outline of First Project	(A) pp. 15-34 (C) pp. 27-120 (D) pp.1-15
3	Representing Knowledge Rule Structures First Project Review	(A) pp. 35- 45 (B) pp. 34-76 (D) pp. 16-29
4	Inference Engine in Insight2+ Drawing Inferences, Parallels to Dynamic Programming	(D) pp. 30-42
5	Incorporation of Uncertainty in Expert System First Project Due and Class-Room Presentation of Projects	(D) pp. 43-108
6	Review of Commercial Languages and Tools Selection of Final Project	(B) pp. 77-174
7	Development of Large Knowledge Systems First Review of Final Project	(B) pp. 175-208
8	Interfacing Expert Systems with Conventional Programs and Databases	(D) pp. 108-118
9	Expert Systems in Civil Engineering, Structures and Geotechnical Second Review of Final Project	(A) pp.49-84
10	Expert Systems in Civil Engineering,Construction, Environmental and Transportation	(A) pp. 85-144
Final	Presentation of Final Projects and Final Exams	

(A) Expert Systems for Civil Engineers: Technology and Application, edited by Mary Lou Maher, ASCE, New York, NY, 1987.
(B) Expert Systems: Artificial Intelligence in Business, P. Harmon and D. King, Wiley Press, New York, NY, 1985.
(C) PROLOG, Programming for Artificial Intelligence, I. Bratko, Addison-Wesley, Menlo Park CA, 1986.
(D) Short Course on Expert Systems for Civil Engineers, Class Notes, Department of Civil Engineering, University of Washington, Seattle, WA, 98195, 1987.

*Weekly homework assignments are due at the beginning of class on Friday, no late homeworks accepted.

TABLE 7-1: Course Syllabus

encouraged to explore PROLOG's descriptive problem solving abilities. Using the logic programming capability of PROLOG, the students are required to understand the differences between prescriptive and descriptive programming and learn simple concepts of predicate logic. The homework assignment for the first week includes the development of a simple PROLOG program and an assignment of illustrating where expert systems are appropriate in civil engineering.

The second week of the quarter the topic of recursion is described, again with the use of the PROLOG language. Simple sorting, ranking, and search procedures using PROLOG are implemented. The discussion of PROLOG is concluded with a simple example of an expert system developed with PROLOG.

The second week also introduces the expert systems shell INSIGHT2+. The choice of this software for the course is described elsewhere in this paper. During the lectures, example programs using INSIGHT2+ are demonstrated and the students are made familiar with the systems basic operation and unique programming features. The assignments for this week require the students to submit a detailed description of their mid-quarter project. This assignment is made early in the quarter to ensure that the students devote sufficient attention to the project. The students are also required to develop a simple expert system (less than ten rules) using INSIGHT2+.

The third week begins a more in depth analysis of the characteristics of expert systems. Lectures are devoted to the process of representing knowledge in the form of rules. The limitations of rules as a means of conveying knowledge are evaluated. The specific rule structure of INSIGHT2+ is discussed, after the general topic of rule development is covered. The homework assignment requires the student to develop several simple expert systems with different types of rule structures. Emphasis is placed on the differences between systems composed of many rules versus systems composed of a small number of complex rules. The students are also required to provide a status report on their first project.

During the fourth week the primary topic is the logic contained in inference engines. Both PROLOG examples and INSIGHT2+ are used to demonstrate the various search techniques that are available. This material is considered important if the student is to develop customized systems in the future or to modify existing systems. INSIGHT2+'s use of a hierarchical goal structure and its relationship to rules are also discussed to demonstrate how inference engines can be controlled. Parallels to Bellman's dynamic programming are also made. The homework assignment requires the students to evaluate the impact of rule order and goal structure on the search procedure used in INSIGHT2+.

The topic of uncertainty management is the primary topic of the fifth week. This topic is viewed as extremely important as it presents one of the major conceptual differences between expert systems and conventional programming. The students are reminded of traditional engineering approaches to decision making such as expected monetary value and Bayes Theorem. They are also introduced to Game Theory because of its close relationship to the MiniMax (or MaxiMin) criteria often used in expert systems. The student is encouraged to challenge the existing paradigms of decision logic incorporated into expert systems and to develop the ability to incorporate their own into their class projects. The homework assignment requires the student to evaluate the appropriateness of various paradigms in several settings.

The fifth week also marks the due date of the first project. The student is reminded that it is his/her first project and that the project will be, by necessity, simple. The students are also encouraged to give careful thought to the classroom presentation of their systems. The presentations are conducted in a semi-formal fashion, requiring the students to present their projects in a fifteen minute period.

The sixth week is devoted to the study of more advanced expert system environments. This is accomplished by literature review and the review of promotional information packages provided by manufacturers. More advanced environments are demonstrated; unfortunately students do not have the opportunity to work in a hands-on setting. It is hoped that the students develop the ability to recognize that the selection of appropriate software requires both a knowledge of the software and the application to which the software is to be placed. The homework assignment requires the student to develop criteria for the evaluation of expert systems software and the application of this criteria to several specific packages and problem settings. Students are also required to submit a description of their final project.

Week seven provides the students with an overview of the development and application of large expert systems. This week is spent reviewing systems outside of the realm of civil engineering. This effort provides the students with a view of the wide range of successful expert systems applications. MYCIN is used as the first case study evaluated. The homework assignment requires the students to review current literature and critique an article describing a relatively new expert system outside of civil engineering. The students must also provide a status report of their second project.

The eighth week examines the use of expert systems with conventional programs and databases. Throughout the course, the students are encouraged to view the emerging field of expert systems as a complement to conventional engineering programming. During this week the students are presented examples of the interfacing of these programming approaches. Again, INSIGHT2+ is used as a prototype. The homework assignment requires the students to integrate a model written in FORTRAN, a PASCAL graphics routine, and disk reads and writes into an expert system.

The ninth and tenth weeks are devoted to the application of expert systems to civil engineering projects. Students are required to review recent literature and report on expert systems applications in their areas of interest. The growing literature of successful applications reported in the ASCE journals are used as a point from which the students begin their search. During these two weeks the students are required to report during class on projects which they have identified.

The final projects are due during finals week. Again, the students make a fifteen minute presentation to describe their results. Written documentation of their efforts are required and a professional presentation of this information is essential for a passing grade in the course. A final examination is also given which covers both general concepts of expert systems and the specific application of INSIGHT2+.

6 HARDWARE AND SOFTWARE CONSIDERATIONS

The Department of Civil Engineering at the University of Washington is fortunate to have excellent computer laboratories. Having participated in a major donation program sponsored by IBM, the department has two computer laboratories with a total of forty IBM/AT machines. The hardware configuration is typically 80286 processors with 80287 math co-processors, 640 K ram memory, and 30 megabyte hard disks. These machines are connected to all mainframes on campus as well as being served by a local token ring network. The laboratories are equipped with a variety of output devices including plotters, line-printers, dot-matrix printers, daisy-wheel printers, and laser printers. The primary displays on these machines are IBM Enhanced Color Displays driven with Enhanced Graphic Adapters. A number of IBM Professional Color Monitors are also available. The equipment in these laboratories is complimented with a fairly wide range of mainframe support on campus including CDC's, VAX's, and IBM's. The IBM donation also provided AT and System 2 computers for a majority of the faculty making it possible for them to conveniently develop software specifically for their classes. General software available on the computers includes programming languages (FORTRAN, PASCAL, C), spreadsheets, database managers, word processors, linear programming packages, CAD facilities, and matrix analysis programs.

The expert system software currently available includes INSIGHT2+, LEVEL 5, and VP-EXPERT. LEVEL 5 is an updated version of INSIGHT2+. INSIGHT2+ is the primary software package used for the expert systems course. It was first marketed in 1984 and is a relatively sophisticated expert system. The expert system can be divided into four integrated components; an inference engine, a text editor, a compiler, and a Help system. The system is written in TURBO PASCAL. Rules are written in a "Production Rule Language." This language is composed of "IF/THEN" type statements, including AND/OR/ELSE type options which makes it possible to develop complex rule systems. Variables in the rules are classified as simple facts (TRUE or FALSE), numeric facts (numbers), string facts (character arrays), and object-attribute facts (variables that can assume one or more values among a potential list of values).

INSIGHT2+ was chosen for the course for a variety of reasons. Documentation for the software is good, with excellent on-line help. The basic commands and format are simple enough to learn in a short period of time. Because the system is written in TURBO PASCAL, it is possible to interface the program with graphic routines written in PASCAL with moderate ease. The system is sufficiently sophisticated to provide a challenge to the students if they are to learn all of its features during a single quarter. In particular, the system has a complex goal structure that allows the development of hierarchical goals, a forward chaining option that uses cycling, and a variety of features to manage uncertainty and its propagation in the expert system.

The configuration of IBM/AT computers and INSIGHT2+ software has proven to be adequate, if not ideal, for course instruction. As a quarter-long course, the software used must be comparatively easy to use and provide good interface facilities with other programs and languages. Students entering the course are familiar with IBM/AT operation so no review is required.

Consideration has been given to modifying both the hardware and software environments, however, it is likely that the hardware environment will remain IBM/ATs for several years. A laboratory of eight Apollo 3500 and 4500 computers is under development in the department. This addition to the IBM/AT equipment will open an opportunity to explore "higher-end" expert system software and will improve the environment in which the current software operates. Although there is interest among the faculty in incorporating more sophisticated expert systems environments into the class (such as KEE, Knowledge Craft, or NEXPERT-OBJECT), it has been concluded that the learning curve for such systems is too slow for an introductory course in expert systems. The use of such hardware and software will be left to research projects on expert systems and to more advanced courses.

7 STUDENT INVOLVEMENT AND CLASS PROJECTS

A primary benefit to the student in enrolling in the expert systems class is the opportunity to explore the use of expert systems software tools. Although the homework assignments are useful in this process, the class project is the principal manner in which the students are required to become intimately familiar with the advantages and disadvantages of expert systems. Class response to such projects has been excellent. In comparison to similar projects in more traditional engineering optimization classes, the students appear to spend significantly more time on these projects and are more motivated by their own accomplishments. As the same faculty teach both of these classes, the students' enthusiasm can not be explained by varying teaching approaches or effectiveness.

One method used to ensure the success of the class projects is to maintain a relatively high profile for the class projects throughout the quarter by making repeated and sequential assignments related to the class projects. Students are first required to submit a description of their potential projects. This is reviewed by the instructor and either accepted, or the student is required to resubmit another topic. The students are then required to present progress reports on their projects including program listings of progress to that date. The students are also required to make formal presentations of their project to the class. The technique of incremental assignments related to the project is used to discourage the student from procrastinating and ensure that orderly progress is made throughout the quarter. The class room presentation is required to give students both practice in oral and presentation skills. It also increases the incentive felt by the students to complete projects because the presentation allows the quality of the project to be exposed to the student's colleagues as well as the instructor.

Students in the classes have produced projects in a wide variety of civil engineering disciplines. Among the topic areas are septic tank design, simple steel structure design, selection and calibration of water quality models, timber construction design, water supply planning, road design for the US Forest

Service, and multi-objective decision making. Students are required in their projects to develop graphical routines in PASCAL to augment data presentation, and to develop a FORTRAN routine to interface with their expert system.

8 FUTURE DIRECTIONS

Although the course in expert systems is relatively new, students have indicated keen interest and it is likely the course will be expanded in the future. The current effort required by students for the course is greater than the course credit they receive. It would therefore be beneficial to develop a second course. A second course would, ideally, allow some of the material currently taught in the first course to be shifted into the second. This would provide the first course with a more reasonable pace and allow more complete coverage of the material that is taught. It would also provide the students an opportunity to further develop systems and to incorporate knowledge acquisition concepts. It seems appropriate that the students become aware of the difficulties that are associated with knowledge acquisition and recognize the commitment of time associated with formal interviews to gain knowledge. The second course would also allow the students an opportunity to work with more advanced expert systems environments such as KEE.

9 COURSE ASSESSMENT

Course assessment is a topic that provides considerable fuel for debate among faculty members. At least two important criteria should be used for assessment, each addressing significantly different concerns. First, is the material in the course relevant to the training of civil engineers? Although civil engineering is a broad field, a reasonable method for evaluating the value of course material to the conduct of the discipline must be established. At issue is the course content and its relationship to the profession. Second, is the course content presented in an environment which allows students to learn the material and encourage their interest in related fields? Quality of teaching, textbooks, homework assignments, availability of instructional help, and hardware and software support fall into this category.

A reasonable debate exists as to the value of expert systems to the discipline of civil engineering. Although a number of successful applications can be found, expert systems have not replaced more traditional engineering analysis nor are there expectations that they will. However, expert systems do provide an excellent environment in which to teach students the "art of engineering". This "art" is sometimes based less on rigorous analysis and more on defining the limits of our knowledge and supplementing it with expertise, experience, and rules of thumb. Recognition that such information plays a significant role in the practice of engineering is essential in engineering education and expert systems provide an excellent avenue for discussing this process. The study of expert systems also provides a formalized logic for including subjective information into an analysis. If for no other reason, it has been concluded at the University of Washington that the field is of sufficient interest and potential value to civil engineers to offer students the opportunity to study the topic. The demand for a course in expert systems will depend upon the successful application of the technique to typical civil engineering problem solving. If continued success stories in this area are not forthcoming, interest will surely lag.

This general question of the role of expert systems in civil engineering suggests another; that is, what constitutes a Master's thesis or PhD dissertation in the field of expert systems when given by a department of civil engineering? Should the contribution to the field be in the area of civil engineering or is it reasonable for the contribution to be in the field of logic expert systems, software design, or decision-making? This question has not been resolved.

In addressing the question of course presentation and content, the facilities at the University of Washington provide an excellent environment for teaching expert systems development. As stated previously, the hardware and software available are adequate to provide students with all of the tools needed to develop systems of their choosing. Sufficient effort has been placed into the development of the course material to give the students every opportunity to immerse themselves in the subject matter. Initial evaluations indicate that the presentation of material has also been adequate.

10 SUMMARY

The development of any new course in a civil engineering curriculum is a challenge. This is particularly true for courses which introduce new approaches to problem solving based on emerging computer technology. A question of concern is whether the material is appropriate for civil engineers and if a civil engineering department is the most appropriate home for such a course. However, if civil engineering departments are to continue attracting outstanding students at the undergraduate and graduate level, emerging computer technology must be incorporated into both the education of civil engineers and into their workplace.

The course on expert systems taught at the University of Washington has addressed this question by emphasizing the application of the technology to civil engineering problems and by providing students the opportunity to develop applications for topics related to their interests. This approach has allowed students to evaluate the role which subjective information, experience, and rules of thumb play in the practice of civil engineering. It has also provided a formalized procedure for the incorporation of such factors into analysis. Of particular value, the course has required students to consider problems that contain subjective information. This differentiation of problem types is useful in preparing students to consider the solution of "real world" problems in which subjective information often dominates the solution of a problem.

Acknowledgments. The author would like to thank Drs. Brian Mar, Greg Miller, and Dorothy Reed of the Department of Civil Engineering, University of Washington for their comments and suggestions.

11 REFERENCES

Johnston, D.M., and R.N. Palmer, Application of fuzzy decision making: An evaluation, *Journal of Civil Engineering Systems,* Vol. 5, No. 2, June, 1988, pp. 87-92.

Miller, G., R. "A Lisp-Based Object-Oriented Approach to Structural Analysis", *Engineering with Computers*, in press, 1988.

Mar, B.W., and R. N. Palmer, Work stations for systems engineers, accepted for publication in *Systems Practice*, February, 1988.

Palmer, R.N., Editorial and introduction to: Special Issue On Expert Systems, *Journal of Computing in Civil Engineering,* ASCE, Vol. 1,No. 4, October, 1987, pp. 1-2.

Palmer, R. N., and K. J. Holmes, Operational guidance during droughts: An expert system approach, *Journal of Water Resources Planning and Management Division,* ASCE, Vol. 114, No. 6, November, 1988, pp. 647-666.

Palmer, R.N., and B.W. Mar, The automation of civil engineers: Some observations, accepted for publication in *Journal of Management in Engineering*, ASCE, November, 1988.

Palmer, R.N., and B.W. Mar, Expert systems software for civil engineering applications accepted for publication in the *Journal of Civil Engineering Systems*, July, 1988.

Palmer, R.N., and R.M. Tull, Expert system for drought management planning, *Journal of Computing in Civil Engineering,* ASCE, Vol. 1, No. 4, October, 1987, pp. 284-297.

Reed, D.A., "The Use of Bayes Networks in Evaluating Structural Safety," to appear in Civil Engineering Systems, June,1988.

Reed D.A., "Expert Systems in Wind Engineering", to appear in Journal of Wind Engineering and Industrial Aerodynamics, 1988.

Chapter 8
Teaching Expert Systems Techniques at M.I.T.

by D. Sriram, A. M. ASCE

1 INTRODUCTION

The use of computers has led to significant productivity increases in the engineering industry. Most of the computer-aided engineering applications were restricted to algorithmic computations, such as finite element programs and circuit analysis programs. However, a number of problems encountered in engineering are not amenable to purely algorithmic solutions. These problems are often ill-structured [the term *ill-structured* problems is used here to denote problems that do not have a clearly defined algorithmic solution] and an experienced engineer deals with them using his judgment and experience. The knowledge-based systems (KBS) technology, which emerged out of research in artificial intelligence (AI), offers a methodology to solve these ill-structured engineering problems. The emergence of the KBS technology can be viewed as the *knowledge revolution*; in the 17th century we had the industrial revolution and around the first half of the 20th century we had the electronic revolution. Kurzweil in his Distinguished Series Lecture at M.I.T on December 3, 1987 linked the progress of automation to two industrial revolutions: the first industrial revolution leveraged our physical capabilities, whereas the second industrial revolution - the knowledge revolution - should leverage our mental capabilities.

I started teaching the course entitled: *Building Knowledge Based Expert Systems for Engineering Problem Solving*, in Fall 1986, to introduce the KBS technology to engineering students. This course is open to graduate students in all engineering disciplines and is offered every fall; in the Spring semester I offer an undergraduate course entitled: *Engineering Applications of Artificial Intelligence*, which deals with practical aspects of knowledge-based systems, natural language processing and robotics. The graduate course will be part of the core courses for students deciding to major in intelligent systems. The undergraduate course is required for the Engineering, Systems, and Computations option, administrated by the Department of Civil Engineering.

At the beginning of the course, I make the following points about KBES:

1. Knowledge-based systems (KBS) are computer programs that facilitate the incorporation of different kinds of knowledge. In other words the KBS technology is a higher level programming technique than conventional programming methodologies.
2. If these systems (programs) incorporate human expertise then they are called knowledge-based expert systems (KBES).

In the following sections, I will discuss the philosophy, outline, text, and evaluation of the course.

2 PHILOSOPHY

KBES are normally developed by a team which is comprised of a domain expert (sometimes more than one), with specialized knowledge about the application domain, and one or more knowledge engineers, with generalized knowledge about KBES building tools, who extract the expert's knowledge and translates it into the knowledge-based expert system. In this course I take the view that it would be more fruitful if engineers were taught the tools and techniques of KBES.

In general, the philosophy of the course is predicated on the fact that the students' background and professional aspirations lie in engineering and that they are interested in learning how latest computer-based technologies can be applied for engineering problem solving, in particular their domain of interest. Hence, the course assumes that through adequate education and professional experience the student has acquired basic knowledge of some engineering domain (or related discipline). The course is project-oriented and takes the view that in the near future KBES will find wide spread usage among engineers, just as Fortran or C became popular with engineering companies in the 70's.

3 OUTLINE

Currently, the following topics are covered: Overview of Knowledge-Based Expert Systems,

Knowledge Representation, Basic Search Techniques, Advanced Problem Solving Strategies, Blackboard Architectures, Model-based Diagnosis, Inexact Inference Techniques, Analogical Reasoning, Neural Networks, Rule-based Programming (OPS5 and VPEXPERT), Frame-based Programming (KEE & NEXPERT), Logic-based Programming (PROLOG), Project Management of KBES, Survey of Existing Tools, and Several Case Studies, which are normally given by speakers from the industry. The course meets twice every week, with each session lasting 90 minutes. A list of lectures is provided below.

Lecture No.	Topic
1	Course Overview
2	Applied AI: An Overview
3, 4	Search
5, 6	Knowledge Representation: Rules, Frames
7, 8	Advanced Problem Solving: GPS, Forward Chaining, Backward Chaining, Hierarchical Planning
9	Project Management of KBES
10	Advanced Problem Solving: Constraint Management, TMS Inexact Inference Techniques
11, 12	Rule-based Programming: OPS5 VP-EXPERT
13	Blackboard Architectures
14	Object Oriented Programming
15	Frame-based Systems: KEE NEXPERT, KAPPA
16	Mid-term
17	Logic-based Programming: PROLOG
18	KBES for Engineering Design
19	CASE STUDY: DESIGN
20	Case Based Reasoning
21	Model Based Diagnostic Reasoning
22	CASE STUDY: PLANNING
23	Knowledge Acquisition
24	Neural Networks
25, 26	Project Presentations

4 TEXT

I have edited a book entitled: *Computer Aided Engineering: The Knowledge Frontier*, which is used as the text book; the book is yet to be published (a draft copy of the book and associated overheads are available from the Intelligent Engineering Systems Laboratory, M.I.T.).

The book is divided into two volumes. The first volume contains fundamentals of knowledge-based systems, while the second volume contains application papers. Details of the first volume are provided below.

CHAPTER 1 is an overview of knowledge-based expert systems (KBES). The differences between knowledge-based systems and conventional programs are pointed out. The various components of a knowledge-based expert system are briefly described. The derivation-formation spectrum of problem solving is introduced (See section 6 of this paper). A car diagnosis KBES is illustrated. A number of applications, which appear in Volume II of the book, are described. This is followed by a discussion on the process of building KBES.

Problem solving involves search. A search method can be defined as the specification of a behavior by a rational agent to achieve a certain goal. A number of search methods are described in the artificial intelligence (AI) literature. These search methods - called basic search methods - are divided into: *simple search*, *evaluation-based search*, and *game playing*. Simple search methods are Depth-first and breadth-first search techniques. Hill climbing, best-first search, and A* search come under evaluation-based search methods. In CHAPTER 2, a generic search procedure is developed. Basic search methods are described by extending the generic search procedure; these search methods are illustrated with examples drawn from engineering.

CHAPTER 3 is devoted to knowledge representation. Most KBES deal with symbolic knowledge, represented in the form of rules, semantic networks, frames, objects, and causal knowledge structures. Inference networks, goal networks, and dependency networks, which are usual visual aids for representing various knowledge structures, are presented. For engineering applications, there is a need to couple algorithms with symbolic knowledge. A framework for encoding knowledge at various levels is provided. The concept of object-oriented programming is also included.

The primary concern in applying basic search methods for solving engineering problems is the tendency for these methods to lead to combinatorial search spaces, due to the lack of knowledge needed to guide search. To adequately deal with complex engineering problems basic search methods should be augmented with: 1) domain specific knowledge, and 2) problem solving techniques to deal with: a) goals and data, b) multiple levels of abstraction, c) assumptions and justifications, d) inexactness of knowledge and data, e) time, and f) spatial reasoning. CHAPTER 4 deals with advanced problem solving strategies, which address some of the above issues. Forward Chaining, Backward Chaining, GPS, Problem Reduction Hierarchical Planning, Heuristic-Generate-Test, Backtracking, Dependency-Directed Backtracking, Truth Maintenance, Constraint Handling, Organization of KBES are discussed.

In addition to the advanced problem solving techniques, discussed in Chapter 4, there are several problem solving (PS) architectures that could be of considerable use to researchers in engineering, in particular for researchers working in design automation. A few of these PS architectures are: *Blackboard systems, analogical reasoning, neural networks, and SOAR*. The Blackboard architecture provides a framework for: 1) integrating knowledge from several sources, and 2) representing multiple levels of problem decomposition. It uses two basic strategies: 1) divide and conquer, and 2) opportunistic problem solving. The divide and conquer strategy is realized by decomposing the context, which is called a *blackboard*, into several levels depicting the problem solution decomposition, while opportunistic problem solving is achieved by focusing on the parts of the problem that seem promising. There are several variations of the Blackboard architecture. These variations are discussed in CHAPTER 5.

Experienced engineers have the ability to utilize knowledge gained from previous experiences to provide novel solutions to a wide range of problems. Analogical reasoning (AR) involves the use of past experience to solve problems that are similar to problems solved before. A form of analogical reasoning (AR) - called case-based reasoning (CBR) - has considerable potential for developing KBES in engineering design. This technique is illustrated, with a landscape design example, in CHAPTER 6.

Of late, considerable interest is being shown in a different kind of problem solving architecture called *neural networks*, where knowledge is represented in the form of nodes - called cells - and weights which determine the strength of connections between these cells. Inferencing is performed by adjusting the weights, associated with links, between cells. The neural network-based systems can acquire and process knowledge (assuming some basic knowledge exists) through the adjustments of weights associated with links and through reconfiguration of these links. An introduction to neural networks, examples in engineering, and the use of neural networks as KBES frameworks are the topics covered in CHAPTER 7.

In Chapter 1, various stages involved in the development of a KBES are briefly outlined. In CHAPTER 8, we elaborate on these stages with examples drawn mainly from the COMPASS project for which Prerau, who is a co-author on this chapter, served as the project leader. A brief description of COMPASS is provided. This is followed by discussions of the KBES development cycle. Knowledge acquisition methodologies are addressed in detail. Issues in technology transfer are also discussed.

Chapters 9, 10 and 11 are dedicated to descriptions of various programming tools that are used for implementing KBES. The rule-based programming paradigms are discussed in CHAPTER 9: OPS5, KAS, INSIGHT^{tm}, PERSONAL CONSULTANT^{tm}, RULEMASTER^{tm}, and VP-EXPERT^{tm} are described.

CHAPTER 10 is concerned with frame-based and object-oriented paradigms; the tools that are discuused are: SRL, KEE^{tm}, NEXPERT OBJECT^{tm}, and Smalltalk-80. Logic programming - with PROLOG - is the subject of CHAPTER 11.

CHAPTER 12 is a review and comparison of over fifty tools - commercial and academic - for implementing KBES. Guidelines for the selection of an appropriate tool are discussed.

CHAPTER 13 is a comprehensive bibliography of knowledge-based expert systems in engineering; nearly nine hundred papers in Civil, Chemical, Electrical, and Mechanical engineering are cited. In addition, a list of books that are of considerable interest to the engineering community is provided.

APPENDIX A is a brief introduction to data structures. APPENDIX B is a tutorial introduction to inexact inference techniques. The approaches discussed are: PROSPECTOR implementation of the BAYESIAN approach, CERTAINTY theory, FUZZY SET theory, and DEMPSTER-SHAFER theory. A list of AI companies is provided in APPENDIX C.

5 EVALUATION

The grade in the course is based on the following:

1. Assignments on search and knowledge

representation (15 points)
2. Mid-term (20 points)
3. Assignments leading to the final project (25 points)
4. Final Project (40 points)

The project involves the conception, design, implementation and documentation of a KBES in a domain of the student's choice. The project assignments are designed to closely resemble the actual developments of KBES in the industry, i.e., identification, conceptualization, formalization, implementation, and testing stages.

6 TYPICAL PROJECTS

The kinds of problems that are encountered in an engineering cycle can be represented by the *derivation-formation* spectrum. In derivation problems, the problem conditions are posed as parts of a solution description; the possible outcomes exist in the knowledge-base of a knowledge-based expert system (KBES). The inference mechanism (a typical KBES consists of a knowledge-base and an inference mechanism which incorporates various problem solving strategies) uses the knowledge-base to complete the solution. Essentially, the solution to these problems involves the identification of the solution path. In formation problems, problem conditions are given in the form of properties that a solution must satisfy as a whole; an exact solution does not (normally) exist in the knowledge-base, but the inference mechanism can generate the solution by utilizing knowledge in the knowledge-base.

Typical projects spanning the derivation-formation spectrum for the last two years are shown in Figure 8-1. A star (*) indicates that the project was extended to either a M. S. or a Ph. D. thesis. The projects are mostly developed on personal computers (IBM PC or Macintosh) utilizing a variety of software. The software tool, the student's name, and the student's department are shown in parentheses; a description of the software tools can be found in Chapter 12 of the text book.

7 SUGGESTIONS

Based on my experience with teaching this course for three terms, I have the following suggestions to offer.

1. Make sure that the student does not select a project that is very ambitious. This would only lead to frustration and confusion for the student.
2. A few students will change projects after the third or fourth assignment. Don't be surprised if this happens.
3. The students like to use software tools on personal computers.
4. The students should be encouraged to draw inference or goal networks, as a visual means of representing knowledge.
5. If the student's problem is a derivation-type problem then VP-EXPERT^{tm} on an IBM-PC (an educational version is available for a very reasonable price) is highly recommended. For the Macintosh, LEVEL 5^{tm} and Instant-Expert ^{tm} look promising.
6. For students involved in developing intelligent interfaces with traditional CAE programs, KAPPA^{tm} and NEXPERT-OBJECT were found to be useful. The educational version of NEXPERT-OBJECT is slightly more expensive than KAPPA^{tm}. NEXPERT-OBJECT^{tm} requires a 386-based machine with nearly 4 megabytes of memory and an EGA card/monitor for problems of reasonable size. Students working with KAPPA felt that a 286-based machine with 2 megabytes of memory is adequate for their problems; it must be noted, however, that KAPPA is yet to be tried on large problems in the industry.
7. For formation-type problems, OPS5, NEXPERT-OBJECT, KAPPA, PROLOG were utilized in the projects. However, I feel that NEXPERT-OBJECT, KAPPA, or a similar tool that supports an hybrid programming environment is necessary.
8. It is useful to compile project reports into a single volume and make this volume available to students in later terms (The Fall 1988 projects are available in the form of a technical report from the Intelligent Engineering Systems Laboratory).
9. It might be useful to follow up on the course by offering a knowledge engineering workshop, where a team of students work with an industrial sponsor to develop a real world KBES.

Acknowledgments. *Building KBES for Engineering Problem Solving* is based on *Expert Systems in Civil Engineering*, which is taught by Prof. Steven J. Fenves at Carnegie Mellon

DERIVATION

1. Automobile Diagnosis System
   [VP-EXPERT, Eric Green, CCRE]

2. Analysis of High Accident Locations
   [OPS5, Li-Hung Cheng, CFD]

3. Bridge Cracks Diagnosis*
   [PROLOG, Kim Roddis, CFD]

4. Seismic Safety Assessment for Existing Buildings*
   [PC+, Carolina Mijango, CCRE]

5. Selection of a Floor Framing System
   [GEPSE**, Charis Gantes, CFD]

6. Guide to DAMBRD, a program for unsteady open channel flow *
   [KAPPA, Barbara Hayes, WRD]

7. Gross & Individual Evaluation of Nation's Estauries
   [HYPERCARD, Richard Brynes, OE]

8. Selection of Statistical Methods for Seismic Vulnerability
   [OPS5, Vijaya Halabe, CFD]

9. Determination of Liquefaction Potential
   [VP-EXPERT, Antony Awaida, CCRE]

10. Analysis of Computer Dumps
    [GOLDWORKS, Susan Pitts, EE&CS]

11. Selection of Materials in Design
    [FRAMEKIT/FRULEKIT, Albert Wong, CFD]

FORMATION

1. Roofing System Design
   [OPS5, Daniel Berenato, CCRE]

2. Concrete Beam-Column Joint Design*
   [GEPSE**, Attila Banki, CFD]

3. Holiday Train Crew Cancellation
   [GEPSE**, Patrick Little, TSD]

4. Traffic Signal Intersections Capacity Analysis
   [PC+, Simon Lewis, Urban Planning]

5. Configuration of Local Area Networks
   [OPS5, Didier Perdu, EE&CS]

6. Roller Coaster Design Assistant
   [GOLDWORKS, Thomas Maglione, CCRE]

7. Preliminary Design of Hospitals
   [GBB: Blackboard System, Jim Anderson, Architecture]

8. Selection/Design of Engine Performance Parameters*
   [VP-EXPERT, Ken Patton, ME]

9. Seismic Design *
   [KAPPA, Kevin Cheong, CFD]

10. Rail Scheduling Maintenance*
    [GOLDWORKS, Rabi Mishalani, TSD]

11. Mesh Design for BEM *
    [KAPPA, Mitsunobu Ogihara, CFD]

12. Design of Production Facility*
    [Instant-Expert Plus, Takahide Tokawa, ME]

CFD: Constructed Facilities Division (Dept. of Civil Engineering)
TSD: Transportation Systems Division (Dept. of Civil Engineering)
CCRE: Center for Construction Research and Education (Dept. of Civil Engineering)
WRD: Water Resources Division (Dept. of Civil Engineering)
OE: Ocean Engineering
EE &CS: Electrical Engineering & Computer Science
 ** Currently called KAPPA

Figure 8-1: Sample Projects

University. I am grateful to him for his guidance and help. The development of the book for this course was supported through a curriculum development grant provided by the Dean of Engineering at M.I.T. C. Pouangare and N. Groleau served as teaching assistants for the course during Fall 1987 and Fall 1988, respectively. The comments and suggestions made by the reviewers improved the quality of this presentation.

Chapter 9
Teaching Expert Systems Techniques at Purdue University

by Jeff R. Wright, M. ASCE

1 INTRODUCTION

Though opinions vary about degree to which expert systems technology will alter the future of engineering research and practice, the technology has been clearly embraced by all engineering disciplines. Virtually every engineering department of every major engineering school can point to individuals having at least a token interest in this emerging field. While the appropriate home for this new technology seems to be a bone of contention between electrical and industrial engineering departments, educators across all engineering units are searching for appropriate mechanisms for exposing graduate students—again, to varying degrees—to expert systems theory and methods.

For civil engineering students, the appropriate vehicle for this exposure is one or more graduate elective courses having the right mix of theory and application with an emphasis on design, development and evaluation of working systems. This is an impossible task for any single course and not likely to be achieved through the selection of courses across engineering disciplines; technical courses are designed (and approved) to meet disciplinary needs. Consequently, unless a civil engineering program makes a commitment to a technology such as expert systems to the degree of course sequences and dedicated faculty assignments, students frequently struggle in courses for which they may not have the background or are precluded from such offerings.

An alternate approach to this predicament has been taken within the School of Civil Engineering at Purdue University. A course has been designed that attempts to meet the specific needs of that school's graduate students and, at the same time, serves a need not presently met by other campus offerings. The course—entitled "Expert Systems Prototype Development"—serves as an introduction to the technology of expert systems for some students and, at the same time, builds upon previous exposure to the theory of expert systems for some others. The goal of the course is to address this important technology from a software engineering perspective. As such, the focus is on the programming tools available for the design and construction of expert systems rather than on the theory behind their design. Students unfamiliar with the technology are exposed to the methods of expert systems through building prototype systems, while those having previous exposure to the theory from other courses learn how to implement the technology. Because the course emphasizes the integration of existing software methodologies rather than the development of new, related technologies such as database management and graphics systems can also be incorporated into the course. Most important, the course serves to complement other offerings among the engineering schools but within the context of a domain of interest to the target student group.

In contrast to other graduate courses in expert systems (Brown 1987), the principal goal of this course is to acquaint students with the methods of building software prototypes—"...*a strategy for performing (software) requirements determination wherein user needs are extracted, presented, and developed by building a working model of the ultimate system—quickly and in context*" (Boar, 1984, page 7).† The course is designed to focus on a specific engineering domain within which students study past applications and develop working prototypes as term projects. Lectures are presented, and homework problems solved, using an appropriate shell tool, with the choice of development environment for an extensive term project prototype left up to the discretion of the student. Students having prior experience with development languages or other expert systems

---

† Evolving from structured programming methods that emerged in the early 80's, prototyping has attracted both enthusiastic cheers and strong resentment among software engineers across the country (Yourdon 1986, Klinger 1986, Estvanik 1988, Cholawsky 1988).

tools may use those systems with which they are most comfortable.

As indicated above, no single course can meet all the needs of all students and this course is no exception. By design, it is not intended to exhaustively present the theory nor the practice of expert systems. However, in its present form, it seems to meet the needs of three specific groups of students: 1) upper-level undergraduate students or beginning graduate students curious about expert systems technologies as a possible direction for their graduate studies, 2) students with a firm background in the theory of expert systems gained through exposure to other courses in electrical or industrial engineering, management science, computer science, etc. who are designing or conducting research using these tools within the present domain of focus, and 3) senior graduate students (and faculty) who seek exposure to the use of expert systems methods in engineering problem solving as well as a more complete perspective on modern computer hardware and software systems.

The purpose of this paper is to present the overall structure and organization of this course. Two major themes are addressed: 1) Establishment of a prototyping environment, and 2) Course planning and topic sequencing. Domain specific considerations for this course depend on the topic selected for a given semester and will not be addressed. Some thoughts about our experiences with this course over the past three years and expectations for the future conclude the paper.

2 RATIONALE AND COURSE DESIGN

A well defined "domain of focus" is essential to the success of this course; only if the instructor is firmly within his/her realm of expertise can they keep students focused on the techniques and methods of prototype development. While this focus can (and probably should) change each time the course is offered, it is important that both students and instructor have a clear source of reference materials available as to aid in knowledge base development. Next, because the course is fundamentally one of software engineering, a common programming environment—one that allows most major expert systems concepts to be addressed—must be readily available. Thirdly, perhaps more than other graduate engineering courses, a carefully planned schedule must be developed. Before discussing topic selection and sequencing, we turn our attention to the issues of the proper prototype development environment and course scheduling.

2.1 Expert Systems Prototyping

While there is a great deal of disagreement about precisely what is meant by "prototype software development," there is enthusiastic endorsement of the technique as a framework for the development of expert systems, at least within the research community (Boar 1984, Klinger 1986, Estvanik 1988, Dym 1987, Phillips and Sanders 1988, Cholawsky 1988, Harmon and King 1985, Waterman 1985). The consensus is that prototyping is a process by which a working model of the software system is developed without preliminary and exhaustive specification of system function and performance. The designer is free to experiment with alternate design strategies within the environment that he/she is most familiar. The benefits of this approach in contrast to traditional prespecification methods result from the speed with which systems may be developed and the ease with which user (or client) input may be incorporated into the design. The problems resulting from this approach have to do with such things as accountability, design coordination, system development scheduling, etc.; issues important to commercial software firms but of little concern within the classroom. Leaving the precise definition (description) of prototyping to others, the characteristics of this methodology that make it the development framework of choice for this course include:

1. While developing the prototype, one need not be concerned with specific hardware and software requirements. The system designer is free to use systems with which they are familiar.

2. System performance need not be a concern, only system function. Through the development process, designers will be able to identify those efficiency issues that will need to be addressed in the production development phase of the project such as the appropriate choice of hardware and software for the production system.

3. Documentation requirements are not severe. Developers need only provide documentation for their own use rather than for use by end users and programmers.

4. Communication between and among system modules need not be consistent throughout the application. Flexible combinations of both loosely and tightly coupled systems are permitted relaxing unnecessary constraints on system function.

5. Development may begin immediately, without the delays normally required to assemble a pre-specified development platform.

Of particular importance for expert systems development is the flexibility afforded by prototyping methods relative to communication with other programs and systems. This might include such things as links to existing databases, interfaces to graphical display routines, or interaction with some pre-existing model for support analysis. The prototype model of development permits the use of these communication mechanisms—or any combination—as needed by the knowledge engineer (student).

Consider the schematic diagram of a generic prototype presented as Figure 9-1. The configuration shown consists of an expert systems *executive* and three external program elements or modules, each supported by a separate communication mechanism. Using an appropriate shell tool (discussed below), students design and build this executive module containing (at least) a formal knowledge base and a user interface, and that typically requires some formal data inputs. Module 1 represents an external program or analysis package that might be invoked to perform some specific task such as simulation, optimization or numerical analysis. This module may also have its own data requirements. Communication between the executive and Module 1 may occur in one of two ways. First, a data value might be returned directly to the executive through a pipe or filter mechanism. Alternately, information from Module 1 might be written to an external data file and later recalled by the executive for further use. Module 1 is said to be "loosely coupled" in that the specification of module structure and function may be independent from the design of the executive (Cox 1986, DeWilde 1987). Similarly, Module 3 may be employed to perform some monitoring or display function and although specific data format requirements may be necessary, design of the executive system does not depend on these formats. Module 2 on the other hand might consist of library calls or internal functions that make up an integral part of the knowledge base of inference mechanism. This element is said to be "tightly coupled" because its structure, function as well as implementation are integral to the overall function of this prototype.

The specific prototype design will vary from application to application and depend a great deal on the backgrounds and expertise of the student knowledge engineers. Students with little programming background will be limited to developing simple expert systems executive modules and will concentrate on knowledge base structure and design of the user interface. Students with more programming experience or the use of some specific analytical procedures may experiment with loose coupling of support programs and models that will support and communicate with their executive routines. Students with extensive programming systems may wish to explore ways of embedding other applications into the executive program or design internal functions and routines to explore alternate inferencing schemes. The important idea is that the computing environment—both hardware and software—for this course should provide flexibility, and should accommodate all levels of computer and programming experience.

Fortunately, within the resources of most university engineering programs, establishing a flexible computing environment is relatively easy to accomplish. As discussed below, this course "teaches" one expert system development tool and, specifically, not a traditional programming language. There are three reasons for this. First, while languages might be more simple to access—courses are generally not budgeted for software purchases—and tend to be supported by more complete documentation, they are generally more difficult to teach as a secondary topic within the time constraints imposed by the schedule to be discussed in the next section of this paper. Second, the shell tool selected for use typically provides common ground for students with differing backgrounds as in all our experiences thus far, no student has had previous experience with the programming environment selected for use. Lastly, and perhaps most important, students get first-hand exposure to this relatively new, and dramatically changing technology.

The debate over the merits of traditional programming languages vs. expert systems development environments and prototyping tools is broad and without hope of reaching closure (see, for example Barber 1987, McCormack 1985, Tucker 1986). The selection of an appropriate expert system prototyping environment is a topic beyond

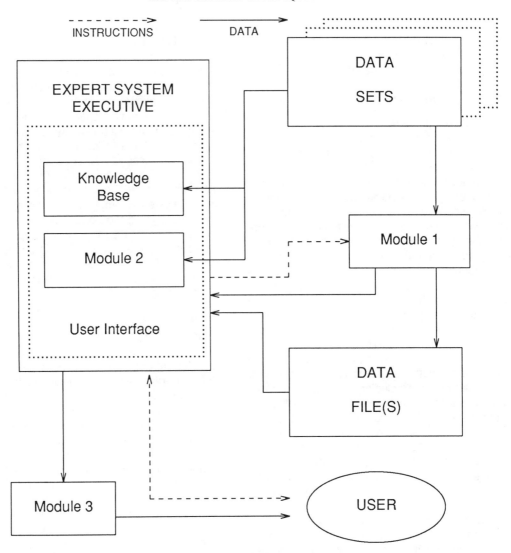

Figure 9-1: Examples of Prototype Cohesion

the scope of this paper and has been adequately addressed elsewhere by others (see, for example, Mullarkey 1987, Engel et al. 1988). For our purposes however, the following considerations are important:

1. The development tool must be easy to learn and use and must have good user documentation. Programmer documentation is less important but may be a concern if a significant number of students have extensive programming experience.

2. The shell must support (at least) both rule and frame-based inference mechanisms. Alternately, two different systems might be used.

3. The shell must provide mechanisms for communication among and between other modules similar to the scheme presented in Figure 9-2, above.

Because roughly 1/3 of the course will be devoted to the syntax of the shell and its use for prototype development, it is important that the instructor(s) be very experienced with the system selected. Furthermore, in the spirit of the goal that this course reflect the state-of-the-art in expert systems technology, the same shell environment need not be used each time the course is offered. Indeed, the author has benefited greatly from his experiences with different tools each time the course has been given with the added benefit of staying current with the technology in a relatively painless manner.

2.2 Course Format and Sequencing of Material

The third ingredient for a successful course in expert systems prototyping is careful planning and sequencing of material, both lectures and program development experiences (homework and projects). In fact, because of the typically wide diversity in the backgrounds of the participating students, we have found that careful planning and sequencing of course materials is probably the most important factor in achieving a satisfying experience for all (or at least most) of the students. If the course is to be more than a simple programming or syntax course, all lectures, readings, homeworks and projects must be designed to support as thorough an analysis of the domain of focus as possible. It is the experience of the author that this is best accomplished by viewing the overall effort of the students falling into three distinct, but related areas: 1) Study of the domain of focus including past expert systems applications as well as sources of knowledge and future needs; 2) Skills in using a shell tool to build knowledge representation schemes and inference procedures, and 3) Design and development of a complete and functioning expert system prototype within the domain of focus using the concepts and principles of expert systems technology. Careful sequencing of these areas of student effort is necessary for a successful course experience.

From our experience, sequencing consistent with the profile presented in Figure 9-2 provides the best integration of these course elements. The vertical axis represents the level of student effort in the course versus time on the horizontal axis. Total student effort is generally expected to be constant throughout the semester as indicated by the dashed horizontal line at the top of the figure. The three curves across the figure indicate effort focused on studying the domain of focus, shell environment syntax and techniques, and project prototype development, respectively.

The first two weeks of the course are allocated to a brief, but thorough introduction to expert systems and applications to the domain of focus. By about the third week of the semester, the emphasis on domain considerations gives way to a rigorous exposure to the syntax of the development environment. As students become comfortable with the methods and capabilities of the software being used, they begin design and evaluation of an actual working prototype expert system that will be presented and discussed formally at the end of the term. Toward the middle of the semester and after becoming conversant with the technology and terminology of expert systems, domain specific issues such as knowledge acquisition and system verification are addressed in more detail. This is followed by more advanced discussions of expert systems concepts. At the halfway point in the semester students should have a working knowledge of expert systems, strong skills in the use of at least one prototyping tool, and a detailed plan for the development of their project prototype. Class lectures during the last third of the course address, for a third time, domain concerns while out-of-class activity focus heavily on the development and testing of individual expert systems prototypes.

Though the emphasis profile discussed above and presented in Figure 9-2 may shift in response to different levels of student experience as well as to different domains, our experience suggest that this general profile has several benefits. First, the question of "domain first or syntax first" is addressed explicitly and "up front." The emphasis on expert systems vocabulary and applications in the domain of focus in the first week may seem a bit overwhelming to students, but it is important in establishing a context for the syntax portion of the course that follows. By mid-semester when attention returns to domain issues, students have become skilled at using the tool of choice and may even be excited about tackling domain-specific issues. Second, following a grounding in syntax and methods, students are able to turn their efforts to actual design and construction of their project prototype. Lastly, a strong finishing focus on domain issues provides serious students with a foundation upon which they might better identify possible directions for their own research efforts. Interestingly, a significant portion of students who have completed this course have subsequently developed thesis research projects related to their course prototypes and, in most cases, using the same development environment presented and used in class.

Achieving the emphasis profile reflected in Figure 9-2 requires careful planning and thoughtful attention to lecture sequences and preparation of homework assignments. Over the past several semesters, a syllabus structure has evolved that supports this approach. Table 9-1 presents a

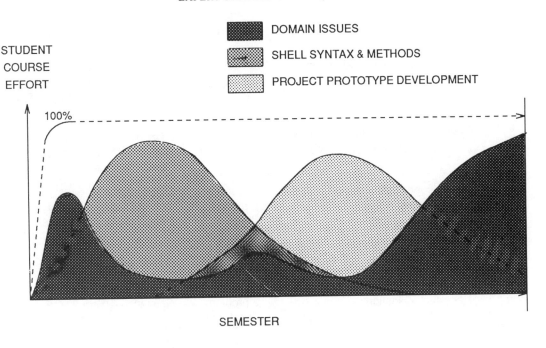

Figure 9-2: Relative Course Emphasis

summary of lecture topics, homework assignments and readings by week for a typical 16-week semester. Together, these three work elements generate a work load for students that follows the profile of Figure 9-2. Student evaluation is based on weekly homework assignments over the first 10 weeks of the course, two mid-term exams, and the project prototype systems accounting for roughly 20, 40 and 40 percent, respectively toward a final grade. Though some domains may dictate special requirements relative to overall timing, the lecture sequence shown in Table 1 has proven effective for most classes.

The first week is a full one. A basic terminology of expert systems is presented following generally the format presented by Maher and Allen (1987). Several examples of prototype development within the domain of focus are introduced and students are asked to read thoroughly the excellent article by Bobrow and Stefik (1986). Additional historical perspective is offered by the chapter by Chandrasekaran (1983) and the piece by Davis (1987). A pragmatic description of functional prototype development is delivered by Weisman (1987). Students are asked to read these materials in the admittedly short period of 1-week, and to bring questions to class for general discussion. In addition, a gentle introduction to the syntax of a rule-based shell makes possible a simple homework assignment designed to acquaint students with their computing environment.

The thrust of the second week is a more detailed look at the domain of focus and a discussion about how and why prototype applications within the domain have been attempted. Special attention is given to the types of development systems that are available as well as the general approaches that have been espoused for system design (Dym 1987, Hayes-Roth et al. 1983). A disciplinary perspective is provided by Fenves (1984) and other domain-specific references. Homework for the week requires students to prepare annotated bibliographies for applications that they are able to find in the library. These bibliographies will be compiled and distributed to the class when complete.

The next four weeks—week 3 through week 6—have a heavy emphasis on the syntax and capabilities of a rule-based shell development environment. Two simple, and one more complicated homework assignments are designed to expose students to a full complement of knowledge base development and inference engine design considerations with lectures and readings selected to provide an appropriate mix

WEEK	TOPIC	Homework	READINGS
1	Introduction to knowledge-based systems and prototyping	Simple 5-rule prototype	*Bobrow & Stefik (1986), Chandrasekaran (1984), Davis (1986), Weisman (1987) Maher & Allen (1987),*
2	Expert systems in domain of focus: summary of major applications	Library Research	*Dym (1987), Fenves (1984), Hayes-Roth et al. (1983), Domain References*
3	Knowledge representation with rules: structure & function	20-rule problem	*Degroff (1987), Wilson (1987), Giboney (1986), Phillips & Sanders (1988)*
4	Logic and flow control with rules: attribute hierarchy; rule ordering	20-rule problem	*Stefik, et al (1983), Software Documentation*
5	Knowledge uncertainty & system validation	Complex rule problem w/externals	*Dong & Wong (1986a,b) Geissman & Schultz (1988)*
6	Systems interfacing & communication: embedding & production	Project Proposal	*Cox (1986; Chapter 2) Software Documentation*
7	Knowledge acquisition issues	EXAM	*Parsaye (1988), Evanson (1988) Waterman (1985; Chapter 14)*
8	Domain design issues	Project Outline	*Domain References*
9	Frame knowledge representation structure & function	Simple-frame problem	*Salzberg (1987) Finin (1986), Marcot (1987)*
10	Classes and attribute inheritance	Complex problem	*Stefik & Bobrow (1985) Software Documentation*
11	Other knowledge representation & Inferencing Mechanisms	Project design	*Thompson & Thompson (1987) Software Documentation*
12	Domain application	EXAM	*Domain References*
13	Domain application	Prototype	*Domain References*
14	Domain application	Prototype	*Domain References*
15	Domain application	Prototype	*Domain References*
16	Class project presentations	Final report	

Table 9-1: Course Syllabus

of theory and application. Articles by Degroff (1987), Wilson (1987), Giboney (1986), and Phillips and Sanders (1988) are selections from the popular literature that provide a more conversational exposure to rule structure and function than found in most text books on expert systems. The articles by Stefik et al. (1983) and Dong and Wong (1986 a & b) are considerably more heady and should be used with caution (discretion?). The chapter by Cox (1986) is particularly well suited for students wishing to learn about the more philosophic issues behind object oriented programming and the piece by Geissman and Schultz (1988) is an interesting discourse on prototype evaluation. It should be noted that as an alternative to some of these references, shell software documentation is often a good source of practical information about systems development.

By the end of the sixth week, students are assumed to have a good command over the shell language as well as a general notion of prototyping methods. Together with a thorough introduction to the concepts of expert systems, each student may be expected to prepare a detailed proposal for a project prototype he/she will build. This proposal should include at a minimum the problem to be solved, source of expertise (use of a human expert is encouraged), specification of a development environment (both hardware and software), and strategy for testing and presenting results. Requiring a preliminary prototype proposal about the end of the third week is useful as are frequent status reports until a full proposal has been prepared.

Just prior to the midpoint of the semester, and while students are working on a moderately complex, multi-week expert system, attention returns to the domain of focus including a sequence of lectures on knowledge acquisition. Articles by Parsaye (1988), Evanson (1988) and the chapter from Waterman (1985) are good references to initiate discussion and, if possible, a guest lecture by a domain expert or group of experts is an extremely useful mechanism at this point in the course. By the end of week 8, each student should have prepared a detailed outline for their term project including a comprehensive literature review.

Development skills are again the major thrust in weeks 9 and 10 with attention turning to frame-based knowledge representation and inferencing (students seem to relate faster to rule structure possibly explaining the preponderance of rule-based expert systems in use). Articles by Salzberg (1987), Finin (1986), and Marcot (1987) are directed toward a general audience while the piece by Stefik and Bobrow (1985) will require more time and thought on the part of most students.

Week 11 is usually one of transition between advanced skills and syntax and domain applications. The instructor may elect to introduce more advanced prototyping topics, to discuss alternate schemes for knowledge representation and reasoning (Thompson and Thompson 1987), or return more directly to discussions about the domain of focus. This period has also been used to address hardware considerations, though this is not generally a formal course topic. In any case, regular homework assignments are dropped in order to provide students more time for prototype development, which should be well under way by this time.

Lectures for the remainder of the semester concentrate on the domain of focus. An attempt is made to uncover all possible applications (good and bad) within that domain. Frequently, this will require a good deal of research and personal communications on the part of the instructor. It is this "home stretch" when one realizes the importance of choosing carefully the domain of focus.

Though formal student evaluation is a matter left to the instructor, it has been our experience that two midterm examinations, one each in weeks 6 and 12 provide necessary "information termination points" for the students. This timing generally avoids the typically heavy 8th and 16th week of the term and is usually appreciated by the clientele. A final exam is not given, but students are expected to make a formal presentation of their project including a real-time demonstration to the instructor. A minimum of 30 minutes is allotted for this important purpose.

3 SUMMARY AND CONCLUSIONS

In most universities, including Purdue, there is still relatively little experience with teaching graduate level expert systems prototyping courses for engineering students. More discussion among engineering faculty is needed to better define the approach, the methods, and the scope of such courses for each domain of focus. This paper is offered as a framework for this discussion.

Clearly, all issues have not been resolved. More work needs to be done, for example, in the development of educational methods and techniques on the topics related to domain knowledge acquisition skills. Although a few sketchy techniques for general knowledge acquisition exist, more refined, domain-specific strategies are still to be developed in such a way that they may be translated to other domains. This is indeed true within the construction engineering and management domain with its need for broad expertise in engineering, business and legal aspects of executing construction project.

The evaluation of development tools is another issue not explicitly addressed in this course. While students acquire strong skills in the use of a specific expert systems prototyping shell, we make no attempt to provide an exhaustive and complete evaluation of the full range of choice nor offer guidelines as to how one might stay current with the technology into their career.

Perhaps the greatest fear of the author is that because this course places a great deal of emphasis on programming, students will become comfortable with syntax to the extent that they confuse prototyping with engineering. The structure and function of the course outlined in this paper is not meant to be offered as one that will, on the basis of this experience alone, produce individuals who can claim to be expert system developers or researchers. A recurring theme of the course needs to be that the techniques presented are founded on a strong theoretical basis, and that students wishing to attain competency as knowledge engineers must delve deeper into the discipline.

The course outlined in this paper has proven an effective vehicle for providing students exposure to solid expert systems prototyping techniques. As might be expected, the course is not an easy one to teach. A great deal of preliminary planning is needed to select and bound the domain to be used as the focus of the course. In addition, the success of the course rests heavily on the practical experience gained by students in the design and development of their term project prototype. Instructors must be prepared to use a heavy boot but hold a tight rein to keep students directed on task. Most important, the instructor must have a thorough knowledge about both the domain of focus and the capabilities and limitations of the shell tool used for concept presentation. Implicit in the goal of this course to use state-of-the-art technology, the faculty person must be prepared to spend long hours at a keyboard. But while the effort required to teach this course is great, our experience has been that the payoffs are more than adequate.

4 REFERENCES

Barber, G.R., LISP vs. C for implementing expert systems, *AI Expert,* 28-31, February 1987.

Boar, B.H., *Application Prototyping: A Requirements Definition Strategy for the 80s,* John Wiley & Sons, New York 1984.

Bobrow, D.B. and M.J. Stefik, Perspectives on artificial intelligence programming, *Science,* Vol. 231, 951-957, February 1986.

Brown, D.C., A graduate-level course in expert systems, *AI Magazine,* 33-39, Fall, 1987.

Chandrasekaran, B., Expert systems: Matching techniques to tasks, Chapter 4 in *Artificial Intelligence Applications for Business,* W. Reitman (ed.), Norwood, New Jersey; Ablex Pub. Corp., 1983.

Cholawsky, E.M., Beating the prototype blues, *AI Expert,* 42-49, December, 1988.

Cox, B.J., *Object Oriented Programming: An Evolutionary Approach,* Addison-Wesley, Boston, 1986.

Davis, R., Knowledge-based systems, *Science,* Vol. 231, 957-963, February, 1986.

Degroff, L., Conventional languages and expert systems, *AI Expert,* 32-36, April, 1987.

DeWilde, G., Developing modular programs, *Computer Language,* 51-55, January, 1987.

Dong, W. and F.S. Wong, From uncertainty to approximate reasoning: part 1: conceptual models and engineering interpretations, *Civil Engineering Systems,* Vol 3., 142-154, September, 1986.

Dong, W. and F.S. Wong, From uncertainty to approximate reasoning: part 2: reasoning with algorithmic rules, *Civil Engineering Systems,* Vol 3., 192-202, December, 1986.

Dym, C.L., Implementation issues in the building of expert systems, Chapter 2 in *Expert Systems for Civil Engineers: Technology and Application,* M.L. Maher, ed., New York: ASCE, 1987.

Engel, B.A., D.D. Jones and J.R. Wright, "Selection of an expert system development tool," *Proceedings of the 1988 Summer Meeting of the American Society of Agricultural Engineers,* Rapid City, South Dakota, 7-pages, June, 1988.

Estvanik, S., Prototyping software: Build one to throw away, *Computer Language,* 71-74, November, 1988.

Evanson, S., How to talk to an expert, *AI Expert,* 36-42, February, 1988.

Fenves, S.J., Artificial intelligence-based methods for infrastructure evaluation and repair, In *Infrastructure: Maintenance and Repair of Public Works* New York Academy of Sciences Report Series Vol. 431, pp. 182-193, 1984.

Finin, T., Understanding frame languages, *AI Expert,* 44-50, November, 1986.

Geissman, J.R. and R.D. Schultz, Verification and validation, *AI Expert,* 26-33, February, 1988.

Giboney, V., Conventional programming and expert systems, *Computer Language,* 53-60, August, 1966.

Harmon, P. and D. King, *Expert Systems: Artificial Intelligence in Business,* New York: John Wiley & Sons, 1985.

Hayes-Roth, F., D.A. Waterman, and D.B. Lenat, An overview of expert systems, Chapter 1 in *Building Expert Systems,* F. Hayes-Roth, D.A. Waterman, and D.B. Lenat (eds.), Reading, Mass: Addison-Wesley, 1983.

Klinger, D.E., Rapid prototyping revisited, *Datamation,* 131-132, October, 1986.

Maher, M.L. and R. Allen, Expert Systems Components, Chapter 1 in *Expert Systems for Civil Engineers: Technology and Application,* M.L. Maher, ed., New York: ASCE, 1987.

Marcot, B., Testing your knowledge base, AI Expert, 42-47, August, 1987.

McCormac, J., The right language for the job, *UNIX Review,* 22-32, September, 1985.

Mullarkey, P.W., Languages and Tools for Building Expert Systems, Chapter 2 in *Expert Systems for Civil Engineers: Technology and Application,* M.L. Maher, ed., New York: ASCE, 1987.

Parsaye, K., Acquiring and verifying knowledge automatically, *AI Expert,* 48-63, May, 1988.

Phillips, J.S. and P. Sanders, First steps in prototyping, *AI Expert,* 64-68, May, 1988.

Salzberg, S., Real world knowledge representation, *AI Expert,* 32-39, August, 1987.

Stefik, M., J. Aikins, R. Blazer, J. Benoit, L. Birnbaum, F. Hayes-Roth, E. Sacerdoti, Basic concepts for building expert systems, Chapter 3 in *Building Expert Systems,* F. Hayes-Roth, D.A. Waterman, and D.B. Lenat (eds.), Reading, Mass: Addison-Wesley, 1983.

Stefik, M. and D.G. Bobrow, Object-oriented programming: Themes and variations, *The AI Magazine,* pp. 40-62, Fall, 1985.

Thompson, B. and B. Thompson, Creating expert systems from examples, *AI Expert,* 21-26, January, 1987.

Tucker, A.B., *Programming Languages,* New York: McGraw-Hill, 1986.

Waterman, D.A., *A Guide to Expert Systems,* Reading, Mass: Addison-Wesley, 1985.

Weisman, R., Six steps to AI-based functional prototyping, *Datamation,* 71-72, August, 1987.

Wilson, L., Rule-based Programming in C, *AI Expert* 15-21, August, 1987.

Yourdon, E., What ever happened to structured analysis?, *Datamation,* 133-138, June, 1986.

Chapter 10
Teaching Expert Systems Short Courses: A Personal Perspective

by Richard N. Palmer, M. ASCE

1 INTRODUCTION

This paper concerns the development and teaching of a short course in Expert Systems in Civil Engineering. Like other short courses, its purpose is somewhat different from that of a course taught at a university during a semester or quarter time period. In addition, the students involved in such a course are very different than those found in a typical undergraduate or graduate course. These differences provide a unique challenge to those teaching the class and demand an approach and philosophy that meet these students' particular needs. This paper describes a short course that has developed at the University of Washington and has been taught twice annually for the past three years for the Systems Engineering Division at the Boeing Aerospace Company. In addition, variations of the course have been taught at the Universidad Politecnica de Catalunya, Barcelona, Spain (June, 1988), the Norwegian Institute for Water Resources, Oslo, Norway (September, 1988), the Institute of Hydraulic Engineering at Dubrovinik, Yugoslavia (June, 1989) and in Seattle, Washington (July, 1989). The course enrollment has averaged approximately 25 participants, with classes as large as 32 and as small as 21.

The purpose of this paper is to present the general approach used to teach short courses on the topic of expert systems applied to civil engineering and to present the difficulties, failures, and (hopefully) the lessons learned in teaching a course in a new computer application field to a diverse audience. The author recognizes that the information provided here will be quickly dated. However, the paper presents one educator's approach to the topic during a period of expanding applications in expert systems to civil engineering problems. The remainder of this paper begins with some comments on teaching short courses in general and expert systems in particular. Next, the objectives of the author's expert systems course are presented. A description of the course's organization follows with the course syllabus and classroom assignments. A description of the text used is given, together with a typical example problem. The paper concludes with a brief critique of the courses and comments about the direction the course will assume in the future.

2 SHORT COURSES

Short courses differ dramatically from other teaching experiences. There are many reasons for this. The primary reason is that participants are not students but are practicing professionals. This difference has an impact on a variety of characteristics of the class participants such as: motivation of students, educational diversity, job diversity, average age, and recent educational experiences.

Short course students typically are very motivated. They have chosen to invest their time and money (or that of their employer) in the pursuit of a new skill or knowledge. They expect the course to provide useful information that will have an immediate impact on their work. Unlike undergraduate students, they are not willing to spend time studying theoretical considerations that may or may not have value in practice. They are seeking the maximum amount of information in the smallest time possible.

Short courses are different from traditional class room teaching in that they require an intense transfer of information, restrict the time available for outside assignments, and dramatically change the classroom dynamics. Short courses are demanding for the student as well as the instructor. In a single class room it is possible to have not only a wide range of educational backgrounds but truly diverse abilities. The job diversity and work experience of the group can create situations in which particular individuals may be at a significant disadvantage in learning the material. If this is not recognized in the initial stages of the course, those individuals may not have the opportunity to participate effectively and benefit from the course. The age of the participants is obviously different than traditional courses. This can affect the amount of time the participants are willing to devote to lectures and the interactions between the instructor and the students.

The time constraints of short courses require an intense transfer of information. This can eliminate some of the natural exchange between student and instructor. Homework and assignments require a somewhat different approach since extensive repetition of examples through homework is not possible and little time exists for outside of class assignments. Assignments and examples must be well considered and directly illustrate a point or conclusion made during the class. There is little time for the student to explore alternative problem formulation and solutions.

The classroom dynamics are also altered. Insufficient time exists to make major adjustments to the unexpected, both for the instructor and the student. If participants prove to be more (or less) knowledgeable than the instructor anticipated, it is very difficult to alter the prepared course outline and presentation. It is also difficult to develop new or alternative problems due to time limitations. Recognizing the time limitations, students demand well organized lectures. They also tend to be less accepting of other students that they perceive are dominating the classroom discussions.

There are a number of inherently positive aspects to short courses. Because the participants are professionals, they bring a rich background of experience into the classroom that can be exploited to the benefit of the other students. Often the instructor becomes one of the students, learning as much from an individual in the class as the class may learn from him. The professionals' interests in practical applications encourage the instructor to concentrate on examples that are related to specific problems rather than generalized situations.

Expert systems short courses are particularly unique for several reasons. Expert systems is a very new topic in civil engineering and participants may be interested in obtaining an overview of the topic. Introductory expert system short courses must present a wide range of material simply to introduce the topic. If hands-on experience with software is desired, the instructor must choose between several available software shells. In the classroom, the instructor is likely to face participants whose computer experience ranges from nonexistent to extensive. If programming is required, students must often unlearn procedural programming techniques to learn descriptive programming or non-procedural rule-base programming.

Civil engineering is a sufficiently broad profession that the very diversity of the students may present a problem. Examples in one area of civil engineering (such as structures) may offer no interests to others (hydraulic engineers). The instructor must also attempt to find those examples of interest to all and to indicate the general applicability of techniques and approaches.

To ensure a successful short course, the instructor must consider all of these factors. This implies special attention to the design of course notes, assignments, course software, and the scheduling of topics. The process begins with a clear definition of course objectives.

3 COURSE OBJECTIVES

The course described in this paper is typically offered under the title of "A Short Course in Expert Systems for Civil Engineers." The course has five primary objectives. These objectives are to provide: 1) an overview of the subject of expert systems, 2) an opportunity for students to use an expert system shell and develop simple expert systems, 3) contrast between expert systems and traditional engineering programming and when each may be effective, 4) examples of areas in civil engineering to which expert systems have been applied, and 5) an appreciation of the difficulties encountered when attempting to develop large, operational expert systems.

The objectives of the course require that computer hardware and software be available that allows the students to see expert systems demonstrated, perform assignments and examples and develop simplified expert systems of their own. This has been accomplished by having one personal computer for every other student. This student/computer ratio has proven ideal, as will be discussed later in this paper.

4 COURSE OVERVIEW

As with all short courses, the selection of material to cover, the extent of the coverage, and class interactions are particularly critical due to the short period of time available for the class. The overview material is essential in setting the proper tone for the course and providing the participants with a clear statement of what they are to learn. As a new technology applied to the field of civil engineering, the topic of expert systems is receiving a remarkable amount of attention. When someone unfamiliar with computers and computerized analysis and decision making reads the literature that has appeared related to expert systems, they can mistakenly conclude that such computer programs will replace all other engineering programming and perhaps engineers themselves. Often the literature presents an extremely optimistic outlook for the potential of expert systems. It is important in the overview portion of the course to remind the students of the inherent difficulties in building models of any complex system, regardless of the use of an expert systems shell. It is also important to emphasize that although shells may make it easier to code knowledge into a computer program, knowledge acquisition is still a very difficult and time consuming process.

5 USE OF EXPERT SYSTEMS SHELL

One way to provide an appreciation among the students of exactly how expert systems operate is to provide them an opportunity to develop simple expert systems of their own. This is accomplished with the use of the expert system shell, Insight 2^+. This software is an earlier version of the Level 5 Expert System Software marketed by Information Builders, Inc. This software was chosen from a number of other "lower end" systems as a training tool because of its ease of use, simple rule base system, and its excellent tutorial and prototype shell that have been made available from Information Builders, Inc. at no cost. This prototype shell is limited in the number of rules that it can incorporate and in the functions that it can perform. The system is sufficiently complex, however, to allow students to develop realistic systems in a short course setting.

A majority of time in the short course is devoted to the use of Insight 2^+. This focus is made not because of the unique characteristics of Insight 2^+ but because the training time, options available, and limitations of this shell are typical of the lower end systems that an engineer might use when first investigating the application of expert systems to a particular project. Lessons learned in logic and model development using this software are easily transferable to other expert system shells.

6 LIKELY CANDIDATES FOR EXPERT SYSTEMS

Only after the student has participated in the development of simple systems can they begin to appreciate the limitations of expert systems and the setting in which they may prove to be useful. The fundamental differences between expert systems and more procedural programming become obvious, as does the potential for combining the two into a single system. Rules for recognizing settings for which expert systems are appropriate are provided in the short course. The simple rules and goal structures that are developed in the students' systems, however, do more to clarify and define characteristics that make a problem appropriate for expert systems than any lecture or description provided by the instructor.

7 LITERATURE REVIEW AND EXAMPLES

To broaden the student's exposure to the potential of expert systems, examples of expert systems developed for civil engineering applications are presented. These presentations have taken several forms in the different expert systems courses that have been offered. When possible, speakers in addition to the primary instructor are brought into the class to present the results of their work. When this approach is not possible, a review of the current literature is provided. This process of describing existing expert systems allows the students to understand where other civil engineers have found useful areas of application of expert systems techniques.

8 DIFFICULTIES IN APPLICATION

It is very important for the students to gain an appreciation of the difficulties that are encountered when attempting to develop large, operational expert systems. A primary need for this appreciation is the rather misleading advertising that occurs describing expert systems shells that virtually construct expert systems by themselves and the claim that only a small amount of training and time is required to become familiar with expert systems shells. It is the experience of this author that students quickly understand the time and training commitment that are required for developing expert systems when they have the opportunity to work with commercial shells and when they observe the complex systems that have been built with them.

In addition, the instructor draws upon his experience in the development of expert systems and the obstacles to successful application of expert system techniques. Emphasis is placed on the knowledge acquisition process and the difficulties that are encountered in obtaining information in an organized and useful fashion from experts.

Instruction in the broad objectives defined above are the primary goals in the short course. It has been observed, however, that fully appreciating the lessons of these objectives can only be gained by coming to grips with the process of developing expert systems. The details of how this is accomplished are provided through the course organization.

9 COURSE ORGANIZATION

To address the objectives previously described, a course syllabus has been developed and is presented in Table 1. The syllabus attempts to address each of the objectives of the course. Although a variety of individual topics are discussed, the material can be divided into four basic areas. The course begins with introductory material on the topic of expert systems. This material also discusses a brief history of artificial intelligence (AI) and the potential of expert systems in civil engineering. Next, the use of commercial software shells (such as Insight 2^+) is discussed and demonstrated. The majority of class and laboratory time is devoted to this topic. Third, a review is presented of existing expert systems shells and expert systems applications to civil engineering. Finally, the students are provided an opportunity to demonstrate their short course efforts by presenting the systems they built during the course.

There are five primary topics in the organization of the course that are of interest: 1) the organization of lectures and assignments, 2) the use of Insight 2^+, 3) the development of classroom assignments, 4) the use of class projects, and 5) the development of student teams.

10 LECTURES AND ASSIGNMENTS

A basic premise of the short courses has been that the students can learn best in a teaching environment that combines the use of lectures and laboratories. This environment allows information to be provided to the students followed by an opportunity to immediately apply the information or technique to an example or problem. This has resulted in a class schedule of alternating lectures and computer laboratory exercises of approximately equal length. This format has proven very successful for a variety of reasons. The computer laboratories often serve as a welcome break from the lectures. Such breaks are essential if the students are to maintain a high level of attention during the lecture sessions. The computer laboratories are the most effective way of demonstrating the concepts discussed in the lectures. The computer laboratories are certainly the only way in which the students will get a clear and complete understanding of the skills required in the development of expert systems.

11 INSIGHT 2^+

The software shell Insight 2^+ is used to demonstrate a number of essential features of expert systems. These features include:

Monday, June 20

9:00 AM	Introduction
10:00 AM	Expert Systems Overview
11:00 AM	Coffee
11:30 AM	Insight 2^+, An Introduction
12:15 PM	Computer Laboratory, Chapter 3 Assignment
1:30 PM	LUNCH
3:30 PM	Goal and Rule Structure
4:30 PM	Coffee
5:00 PM	Computer Laboratory, Chapter 4 Assignment
6:30 PM	ADJOURN

Tuesday, June 21

9:00 AM	Fact Types
10:00 AM	Computer Laboratory, Chapter 5 Assignment
11:00 AM	Coffee
11:30 AM	Confidence Factors and Their Control
12:15 PM	Computer Laboratory, Chapter 6 Assignment
1:30 PM	LUNCH
3:30 PM	EXPAND and DISPLAY Functions, TEXT and ASK
4:30 PM	Coffee
5:00 PM	Computer Laboratory, Chapters 7 and 8 Assignments
6:30 PM	ADJOURN

Wednesday, June 22

9:00 AM	Alternate Lines of Reasoning, Forward Chaining
10:00 AM	Computer Laboratory, Chapters 9 and 10 Assignment
11:00 AM	Coffee
11:30 AM	Numeric Facts and Math Functions
12:15 PM	Computer Laboratory, Chapter 11 Assignment
1:30 PM	LUNCH
3:30 PM	CYCLE/STOP/FORGET/INIT, Forward Chaining with Cycling
4:30 PM	Coffee
5:00 PM	Computer Laboratory, Chapters 12 and 13 Assignment
6:30 PM	ADJOURN

Thursday, June 23

9:00 AM	Application of Expert Systems to Civil Engineering
10:00 AM	Review of Expert Systems Software
11:00 AM	Coffee
11:30 AM	An Expert System for Drought Management (SID) - Case Study
12:15 PM	CESOL (Geotechnical Campaign Description)
1:30 PM	LUNCH
3:30 PM	ADOCA (Structures Descriptive Analysis)
4:30 PM	Coffee
5:00 PM	Student Presentations
6:30 PM	ADJOURN

TEXT: Expert Systems for Civil Engineers, Richard N. Palmer, Class Notes, 1988.

Table 10-1: Course Outline

knowledge structure in expert systems, backward chaining, development of goals and rules, fact types, the combination of information and knowledge using confidence factors, user interfaces and their value, forward chaining, the use of mathematical functions in expert systems, and knowledge cycling. As indicated previously, these topics are generic to expert systems and most expert system shells and the understanding gained using Insight 2^+ is similar to that which would be obtained by using any lower end shell.

12 ASSIGNMENTS AND EXAMPLES

For the short course to demonstrate effectively the topics of interest, emphasis must be placed on the development of good classroom assignments. These assignments include both the review of example problems and the completion of in-class assignments. The diversity of the students presents difficulties in the creation of such example problems and assignments. This difficulty has been overcome through two approaches.

A diskette with example rule bases for simple expert systems is provided to the students. These examples take a simple problem and demonstrate the topics that are discussed in the classroom presentations. The topic of ten of the fifteen examples is the decision of the appropriate clothing and rain-gear that should be taken as one leaves home for the trip to their office. This example was chosen for its simplicity, ease of understanding, and its ability to be made more complex. As the example problem expands during the course, it transforms from being a system based upon three simple rules to being quite complex and considering a variety of weather forecasts with varying degrees of uncertainty. A small portion of an example rule base given to the class is presented in Table 2.

Although such a problem does allow the demonstration of important expert system characteristics and techniques, it does not provide a good setting for discussing civil engineering applications. This is corrected by the use of additional in-class assignments. The approach to the in-class assignments is to require the students to define an area of civil engineering in which they have an expertise and to have them develop an expert system on this subject. In this way, they develop a class project which they will demonstrate to others in the class. The system must be unrealistically simple when the student begins, but they are given the opportunity to make it more and more complex as the class progresses. In each assignment, the student is required to incorporate into their model the topic under discussion. Examples of the in-class assignments are presented in the chapter of this book devoted to assignments.

12 PROJECTS

A class project performs several useful goals. It provides the students a point of focus and a reason to maintain enthusiasm in their work. For many, the requirement to present the results of a project to their classmates greatly increases their motivation in the short course. The project also requires the student to consider carefully how a "real" expert system is constructed. It is only when they begin to consider the difficulties presented by the project that they begin to appreciate the effort required. A project also provides the students a source of continuity in the class. Rather than progressing from one unrelated assignment to another, they can focus their attention on one project. Finally, a project can become a source of satisfaction in actually completing a useful piece of work.

14 TEAMS

Because the projects can become large, the students are encouraged to form teams at the beginning of the class and work on projects in groups of two or more. This teaming of students is encouraged for several reasons. In the beginning of the class, the students may exhibit large differences in their confidence in using the computers and expert systems software. It is important to overcome this hesitancy by developing comradery in the groups. There is a large benefit in teaming a student familiar with PCs with one who is just learning about their use.

Teams are also useful in projects because considerable work is required during the short course and a single student simply may not have the time to finish a project. Teaming the students

!
RULE Outside temperature hot
IF   Outside temperature in degrees fahrenheit > = 75
THEN The temperature outside is hot CONFIDENCE 100
!
RULE Outside temperature warm
IF   Outside temperature in degrees fahrenheit < 75
AND Outside temperature in degrees fahrenheit > = 55
THEN The temperature outside is warm CONFIDENCE 100
!
RULE Outside temperature cold
IF   Outside temperature in degrees fahrenheit < 55
AND Outside temperature in degrees fahrenheit > 32
THEN The temperature outside is cold CONFIDENCE 100
!
RULE Outside temperature freezing
IF   Outside temperature in degrees fahrenheit < = 32
THEN The temperature outside is below freezing
CONFIDENCE 100
!
RULE For nimbocumulus clouds - rain
IF Cloud cover IS nimbocumulus
AND The temperature outside is hot
OR   The temperature outside is warm
OR   The temperature outside is cold
THEN Forecast is rain expected CONFIDENCE 80
!
RULE For nimbocumulus clouds - snow
IF Cloud cover IS nimbocumulus
AND The temperature outside is below freezing
THEN Forecast is snow expected CONFIDENCE 70
!
RULE For cirrus clouds
IF Cloud cover IS cirrus
AND The barometer IS rising
THEN Forecast is rain not expected CONFIDENCE 80
!
RULE For cirrus clouds
IF Cloud cover IS cirrus
AND The barometer IS steady
THEN Forecast is rain not expected CONFIDENCE 70
!
RULE For cirrus clouds - rain
IF Cloud cover IS cirrus
AND DISPLAY Barometer

AND The barometer IS falling
AND The temperature outside is hot
OR   The temperature outside is warm
OR   The temperature outside is cold
THEN Forecast is rain expected CONFIDENCE 60
!
RULE For cirrus clouds - snow
IF Cloud cover IS cirrus
AND The barometer IS falling
AND The temperature outside is below freezing
THEN Forecast is snow expected CONFIDENCE 60
!
RULE For altostratus clouds - rain
IF Cloud cover IS altostratus
AND The barometer IS steady
AND The temperature outside is hot
OR   The temperature outside is warm
OR   The temperature outside is cold
THEN Forecast is rain expected CONFIDENCE 65
!
RULE For altostratus clouds - rain
IF Cloud cover IS altostratus
AND DISPLAY Barometer
AND The barometer IS falling
AND The temperature outside is hot
OR   The temperature outside is warm
OR   The temperature outside is cold
THEN Forecast is rain expected CONFIDENCE 75
!
RULE For altostratus clouds - snow
IF Cloud cover IS altostratus
AND The temperature outside is below freezing
AND The barometer IS falling
OR   The barometer IS steady
THEN Forecast is snow expected CONFIDENCE 65
!
RULE For altostratus clouds - clear
IF Cloud cover IS altostratus
AND The barometer IS rising
THEN Forecast is rain not expected CONFIDENCE 60
!
RULE For no cloud cover
IF Cloud cover IS little or no cloud cover
THEN The sky is clear CONFIDENCE 100

TABLE 10-2: Portion of Example Knowledge Base

CHAPTER 1 - INTRODUCTION	1
CHAPTER 2 - EXPERT SYSTEMS: AN OVERVIEW	3
CHAPTER 3 - INSIGHT 2^+: AN EXPERT SYSTEM SHELL	16
CHAPTER 4 - GOAL OUTLINES AND RULE STRUCTURE	23
CHAPTER 5 - PRODUCTION RULE LANGUAGE FACT TYPES	30
CHAPTER 6 - CONFIDENCE	43
CHAPTER 7 - EXPAND AND DISPLAY FUNCTIONS	56
CHAPTER 8 - TEXT AND ASK FUNCTIONS	71
CHAPTER 9 - ALTERNATE LINES OF REASONING	77
CHAPTER 10 - FORWARD CHAINING KNOWLEDGE BASES	83
CHAPTER 11 - NUMERIC FACTS AND MATH FUNCTIONS	90
CHAPTER 12 - CYCLE/STOP/FORGET/INIT FUNCTIONS	96
CHAPTER 13 - FORWARD CHAINING USING CYCLE	103
CHAPTER 14 - DOS ACCESS	109
CHAPTER 15 - CLOSURE	113
REFERENCES	114
APPENDIX A - INSIGHT 2^+ TEXT EDITOR	115
APPENDIX B - DATA TRANSFER	117

TABLE 10-3: TABLE OF CONTENTS OF SHORT COURSE NOTES

allows a more effective use of time and encourages all of the students to perform to the best of their abilities.

15 TEXT AND REFERENCES

The nature of a short course makes readings outside of class difficult. For this reason, a set of notes has been developed for student reference during class and the limited time they may have outside of class. The lectures from the course closely parallel these notes so the students can find supplemental material and alternative explanations for any material in the class. The table of contents for the notes is presented in Table 3. The notes contain extensive references that the student can pursue. Similar to the lectures, the notes devote a significant amount of detail to the operation of Insight 2^+.

16 COURSE CRITIQUE

The student response to this course has been very positive. Students have completed the course with an increased awareness of the potential of expert systems, an understanding of their limitations and restrictions, an appreciation of existing expert system shells, and some knowledge of the application of expert systems to the field of civil engineering.

Perhaps the most important information gained from the short course is a new understanding of how complex decisions can be modeled in an engineering setting. Most participants in the short courses have had experience in the use of numerical models. These individuals often have encountered situations in which insufficient data existed to justify the use of complex numerical models or situations in which the essential aspects of the problem to be solved could not be quantified. They were previously left with no alternative but to use traditional numerical models to aid them in making decisions. Participants in the short course learn that alternative approaches do exist and that there is a strong potential for their use in civil engineering problem solving.

17 FUTURE DIRECTIONS

To remain valuable, a short course such as this must continually be modified to reflect the changes in the field. These changes are of two primary types: 1) changes in software and hardware, and 2) new applications of expert systems to civil engineering. The form of the changes to hardware and software are difficult to predict precisely. However, it can be easily predicted that changes will occur and that faster and better machines will appear at smaller costs per computational unit. Perhaps the most significant short-term change is in the introduction of the 80386 processors (and soon the 80486) to personal computers. High-end expert system software that before only ran on UNIX based microcomputers is now available for 386-based personal computers. Such hardware and software makes available more complex and useful shells and programming languages to a much larger number of professionals. These hardware and software changes must be reflected in the software reviewed during the course and in the computers that are used.

The increased application of expert system technology to civil engineering problems will also provide a valuable research base upon which new and unforeseen applications will emerge. The application of a new technology, such as expert systems, requires the combined efforts of many professionals. Although some of these efforts will indicate only where expert systems are not appropriate or explore approaches that do not prove to be useful, other efforts will prove valuable in pointing the direction of successful applications of expert systems in civil engineering. These effort must be recognized and incorporated into the course to provide the students with a clear indication of the role that expert systems can play in the professional practice of civil engineering.

Chapter 11
The ASCE Short Course on Expert Systems

by Mary Lou Maher, A.M. ASCE

1 INTRODUCTION

The American Society of Civil Engineers has been offering a short course on expert systems since 1986. The course has been organized and taught by the members of the ASCE Expert Systems committee. In two years, the course has been offered five times with an enrollment ranging from 7 to 30 students. The format of the course has varied from a one day lecture to a two day lecture/workshop. The one day lecture includes presentation of the fundamentals of expert systems and examples of civil engineering applications. The two day format includes a hands-on workshop for the second day, during which the students develop a small rule based expert system using a PC based shell.

The students are given a set of course notes and in some cases, a book and/or software. The set of course notes includes a copy of the complete set of transparencies used by the instructors, one or more papers relevant to the course content, and selected bibliographies. The publications distributed to the students include:
1. *Expert Systems for Civil Engineers: Technology and Application*, M.L. Maher (Editor), ASCE, 1986.
2. "What is an Expert System?" by Steven Fenves in *Expert Systems in Civil Engineering*, Kostem and Maher (Editors), ASCE, 1986.
3. "Tools and Techniques for Knowledge Based Expert Systems for Engineering Design" by M.L. Maher, D. Sriram, and S.J. Fenves, *Advances in Engineering Software*, 6(4), 1984.
4. "Engineering Tools for Building Knowledge-Based Systems on Microsystems" by M.R. Wigan, *Microcomputers in Civil Engineering*, 1, 1986.

The book, listed as 1. above, was prepared by the ASCE Expert Systems committee as a result of a nationwide survey of Civil Engineers actively developing expert systems. For the two day course, in addition to the course notes, the educational version of VPExpert was distributed to the students for use during the workshop and to take with them. VPExpert was selected because of low cost and features suitable for developing engineering software (e.g. access to spreadsheets and databases).

2 COURSE OBJECTIVES

The major objective of this course is to present the fundamental concepts associated with current expert system techniques for a civil engineering audience. This objective is identified as satisfying a need for many civil engineers to quickly gain an appreciation for the potential of this new technology in their own work and to identify appropriate actions to take advantage of new computer tools related to expert systems. The course meets this objective by presenting expert system techniques and applications, referring to civil engineering issues when appropriate.

The course is intended for professional and academic civil engineers. The prerequisites for the course include an interest in expert systems and some knowledge of computing practices. The students are not expected to be computer specialists or have prior knowledge of expert system techniques. The course is intended for the engineer that would benefit from an exposure to expert system techniques before committing resources to the use of this technology. The material presented in this course does not distinguish between the use of expert systems for commercial applications or research.

3 COURSE CONTENT

The course is presented in two formats: (1) a one-day lecture or, (2) a one-day lecture followed by a workshop on the second day. The one-day course is oriented towards the presentation of general concepts and civil engineering applications. The two-day course covers the general concepts more quickly and moves on to development issues and preparation for the use of an expert system shell during the second day.

The content of the course has evolved in response to the feedback from the students. The comments form the students included: identify civil engineering applications early; describe a few selected applications rather than an overview of many; and use civil engineering examples when illustrating the concepts and techniques. Overall, the feedback from the students on course content was positive.

4 ONE DAY COURSE

The one-day course is presented as follows.

8:30 - 9:00	Introduction, Agenda, Course Objectives
9:00 - 9:30	What is a Knowledge-Based Expert System?
9:30 - 10:00	Discussion of Civil Engineering Applications
10:00 - 10:30	Demonstration of an Expert System
10:30 - 10:45	Break
10:45 - 11:30	Knowledge Representation Strategies
11:30 - 12:15	Problem Solving Strategies
12:15 - 1:30	Lunch
1:30 - 2:15	Tools for Building KBES
2:15 - 2:45	KBES Development Process
2:45 - 3:00	Break
3:00 - 3:30	Reasoning with Uncertainty
3:30 - 4:00	Issues of Integration of KBES with Conventional Software
4:00 - 4:30	Questions

The first part of the course is designed to familiarize the students with expert systems, their application in civil engineering, and illustrate how expert systems interact with the user. The presentation on "What is an Expert System?" makes a distinction between conventional programming and expert system techniques, provides a brief history of this technology, and comments on the implications of applying this technology to civil engineering practice. The civil engineering applications lecture focuses on the use of expert system technology in the different areas such as structural, construction, geotechnical, transportation and environmental engineering. The different uses of expert systems, such as stand alone consultants, pre- and post-processors to conventional programs, and integration catalysts, are also presented and illustrated. The demonstration of an expert system indicates both the interactive and explanation capabilities of some expert system shells and the specialized knowledge expected by the user of the resulting expert system.

The second part of the course focuses on two fundamental concepts associated with expert systems techniques: knowledge representation and problem solving strategies. The knowledge representation paradigms presented include: rules, frames, object oriented programming, logic programming, semantic nets, and hybrid representations. This presentation stresses the declarative representation of knowledge, drawing on examples from the civil engineering domain. The presentation on problem solving strategies starts with a discussion of forward chaining and backward chaining, using rules and inference networks for illustration. Other strategies presented include search techniques used in Artificial Intelligence as well as the control strategies used in agenda based systems and blackboard architectures.

The third part of the course looks at the more practical issues of developing an expert system. First, the tools available are discussed with respect to generality vs ease of use. The general purpose programming languages presented include C, Lisp, and Prolog. The other extreme is ease of use, including the PC based shells which, in many cases, provide the developer with a user interface, a knowledge acquisition facility, and built-in ability for using certainty factors. A few of the more common tools are presented and illustrated. In the middle of the generality vs ease of use spectrum lie hybrid environments for building expert systems. These tools provide generality through a variety of representation paradigms but usually require more effort to produce a working expert system. The environments presented include Knowledge Craft, ART, and KEE. The tools are discussed collectively in terms of their direct cost and their indirect cost in time required to produce a working and maintainable expert system. The development process is introduced after the discussion on tools. The development process is described as being very similar to any other software project with special emphasis on the role and required participation of a domain expert.

The final portion of the course deals with two special issues of interest to engineers: reasoning with uncertainty and integration. The presentation of reasoning with uncertainty gives an overview of the different methods used to quantify uncertainty in expert systems, including Bayesian inference, Dempster-Shafer Theory, and Confirmation Theory. Examples of the use of these methods are given as well as qualifications on their use. The issue of integration identifies the need for integration and some strategies for achieving integration. The more common applications involving integration are used for illustration, including combining expert systems with CAD programs, and expert systems with analysis programs. Alternative architectures for integration are presented.

5 TWO DAY LECTURE/WORKSHOP

The two-day course is presented as follows.

DAY 1

9:00 - 9:30	Introduction and Agenda
9:30 - 10:00	What is a Expert System?
10:00 - 10:30	Demonstration of an Expert System
10:30 - 10:45	Break
10:45 - 11:30	Knowledge Representation
11:30 - 12:15	Problem Solving Strategies
12:15 - 1:30	Lunch
1:30 - 2:00	Reasoning with Uncertainty
2:00 - 2:30	Integration Issues
2:30 - 2:45	Break
2:45 - 3:30	Tools
3:30 - 4:00	Basics of VPExpert
4:00 - 4:30	Questions to Panel

DAY 2

9:00 - 9:30	Introduction and Agenda
9:30 - 10:15	Development Process
10:15 - 10:30	Break
10:30 - 12:00	Small Groups; Task Selection, A Few Rules
12:00 - 1:00	Lunch
1:00 - 3:00	Development of ES in Small Groups
3:00 - 4:30	Presentations and Discussion
4:30 - 4:45	Evaluation of Workshop

The first day of the two-day course is similar in content to the one-day lecture course. The end of the day includes an overview of VPExpert, the PC based expert system shell the students use on the second day. The students are given a short tutorial with references to the portions of the manual to be read before the next day.

The second day begins with a presentation on the development of expert systems. The development process is presented along with knowledge acquisition techniques and life cycle concerns. The remainder of the day is occupied with the development of an expert system. The students are divided into groups of 3 or 4 according to areas of interest. Within the group, they decide on a particular problem area in which one of them has expertise and they develop a small knowledge base. The projects developed include conceptual design of bridges, estimating soil properties, and construction cost estimating. At the end of the day, the individual projects are presented to the group.

6 SUMMARY

The expert systems course has been successful in providing introductory and application oriented information. The students are generally pleased with the content and presentation of the material. According to the comments from the students, the two-day course was more satisfying; the combination of lecture and workshop provided a better understanding of the technology.

Chapter 12
Text and Reference Books on Expert Systems and Artificial Intelligence

by James H. Garrett, Jr., A.M. ASCE

1 INTRODUCTION

Several years ago, there were few books available that addressed knowledge-based system techniques and could also be used within the classroom. There are now a wide variety of texts ranging from those for introductory courses on the subject to those for use in advanced courses and as references. This chapter is intended to be an aid to an instructor in selecting appropriate course materials for a course on knowledge-based expert systems.

This assistance in selecting a text is provided in two parts. First, several texts covering expert systems and artificial intelligence are briefly described with respect to the following criteria: level, intended audience, and quality of presentation. Following these brief descriptions, a larger, though still incomplete, list of texts and references that address the subjects of knowledge-based expert systems and artificial intelligence is presented.

2 EXPERT SYSTEM TEXTS AND REFERENCES

The following section briefly describes several textbooks covering the subject of knowledge-based experts systems. Each description, presented in a frame-like structure, includes: a few comments about the quality of presentation (e.g., clarity, comprehensibility, good examples and illustrations, etc.), the level of the text (introductory, intermediate, advanced), the intended audience (e.g., computer scientists, engineers, general, etc.), and a listing of the table of contents. The comments expressed are solely those of the author. If two descriptors appear in the level and intended audience slots, it indicates that the text lies somewhere in between, but closer to that listed first.

A Practical Guide to Designing Expert Systems
S. Weiss and C. Kulikowski
Rowman and Allenheld, 1984

COMMENTS:

The scope of this book is somewhat narrow for an introductory text. The authors themselves acknowledge that the book only addresses a well defined class of expert systems, namely "those that solve classification problems using a rule-based approach". However, the text does address the very important subject of system evaluation and verification in more depth than most introductory texts.

LEVEL: Introductory

INTENDED AUDIENCE: General

CONTENTS:

Expert Problem-Solving and Consultation
Reasoning Methods for Expert Systems
Historical Overview of Applications of
   Expert Systems
Designing an Expert System
How to Build a Practical System
Testing and Evaluating an Expert System
The Future of Expert Systems

Artificial Intelligence for Micro-computers - The Guide for Business Decisionmakers
M. Williamson
Brady, 1986

COMMENTS:

This book is intended to provide a businessman a quick introduction to the concepts and jargon of artificial intelligence and, more specifically, knowledge-based systems. It provides brief descriptions of several PC-based software tools, which might be useful for someone trying to select only a PC-based expert system shell. However, this type of information is usually included in other texts that describe software tools for all platforms, not just the PC. The author has done a good job of avoiding jargon and explaining in simple words some of the concepts of artificial intelligence, but in the process, may have oversimplified the technology.

LEVEL: Introductory

INTENDED AUDIENCE: Business / General

CONTENTS:

Artificial Intelligence: The Story Til Now
Natural Language Inquiry Systems
Knowledge-Based and Expert Systems
Starting From Scratch: AI Programming
 Languages
Natural Language Inquiry Systems
Decision Analysis Systems –
 Product Reviews
Expert System Shells – Product Reviews
AI Programming Languages
What's Next?
A Short Course in Computer Literacy
Glossary of AI Terminology
Additional Sources of Information

A Guide to Expert Systems
D. A. Waterman
Addison-Wesley, 1986

COMMENTS:

The presentation of the structure, components, and uses of expert systems is clear and understandable, but somewhat superficial. The book provides very lengthy catalogs and bibliographies of expert systems (subdivided according to the domains of these systems) and expert system tools. Although not complete, which is recognized by the author, these lists are extremely useful for discussing the range and depth of expert system applications.

LEVEL: Introductory

INTENDED AUDIENCE: General

CONTENTS:
What Are Expert Systems?
What Good are Expert Systems?
How are Expert Systems Organized?
How do Expert Systems Differ from
 Conventional Programs?
What Have Expert Systems been used for?
PROSPECTOR: An Expert System at Work
Knowledge Representation in Expert
 Systems
The Nature of Expert System Tools
Stages in the Development of Expert
 System Tools
Examples of Knowledge Engineering
 Languages
Will Expert Systems Work for my Problem?
Building an Expert System,
Choosing a Tool for Building Expert
 Systems
Acquiring Knowledge from the Experts
An Example of the Expert System Building
 Process
Difficulties in Developing an Expert
 System
Common Pitfalls in Planning an
 Expert System
Pitfalls in Dealing with the Domain Expert Pitfalls
 During the Development Process
Where is Expert System Work Being Done? How
are Expert Systems Faring in the
 Commercial Marketplace?
What's Next for Expert Systems?
Sources of Additional Information about
 Expert Systems.
Index for Expert Systems
Catalog of Expert Systems
Bibliography of Expert Systems
Index for Expert System Tools
Catalog of Expert System Tools
Bibliography of Expert System Tools
Companies Engaged in Expert Systems
 Work

Building Expert Systems
F. Hayes-Roth, D. A. Waterman,
and D. B. Lenat, eds.
Addison-Wesley, 1983

COMMENTS:

This text was one of the first general texts on this subject available. It provides a more advanced coverage of the concepts of knowledge-based systems. Because it is composed of contributions from many leaders in the field of artificial intelligence, the presentations of the technology go into much more technical depth than most of the introductory texts. However, because of the multitude of authors, continuity is lacking. Hence, this text is better suited as a reference or an intermediate course text than as an introductory text.

LEVEL: Intermediate

INTENDED AUDIENCE: General

CONTENTS:

An Overview of Expert Systems
What are Expert Systems?
Basic Concepts for Building Expert Systems
The Architecture of Expert Systems
Constructing an Expert System
An Investigation of Tools for Building an Expert System
Reasoning about Reasoning
Evaluation of Expert Systems: Issues and Case Studies
Languages and Tools for Knowledge Engineering
A Typical Problem for Expert Systems

Building Expert Systems – A Tutorial

J. Martin and S. W. Oxman
Prentice-Hall, 1988

COMMENTS:

This text provides several detailed and clearly written case studies on such systems as STEAMER (for assisting naval engineers in controlling steam turbine systems) and DRILLING ADVISOR (advises drill operators on techniques for freeing stuck bits. These case studies provide the depth of coverage not found in most of the current introductory expert system texts.

LEVEL: Introductory

INTENDED AUDIENCE: General

CONTENTS:

What is an Expert System?
The Future of Expert Systems
Terminology
Major Opportunities for Expert System Development
Task Suitability
Building Expert Systems
The Expert System Life Cycle
Expert System Construction Requirements
Languages
Expert System Shells
Commercial Expert System Shells
Hardware for Expert Systems
Development of a Large Expert System
Development of a Personal Computer Expert System
How to Select the Right Tools
The Future of Expert System Technology

Computer-Based Medical Consultations: MYCIN

E. H. Shortliffe
Elsevier, 1976

COMMENTS:

This reference describes the most cited and publicized rule-based expert system, MYCIN, which assists physicians in diagnosing infectious blood diseases and selecting appropriate therapies. The text describes, in detail, the design criteria, architecture, knowledge representation strategies, and the inexact reasoning strategy used in MYCIN.

LEVEL: Intermediate

INTENDED AUDIENCE: General

CONTENTS:

Introduction
Design Considerations for MYCIN
Consultation System
Model of Inexact Reasoning
Explanation System
Future Directions for MYCIN
Conclusions

Crafting Knowledge-Based Systems - Expert Systems Made Realistic
J. R. Walters and N. R. Nielson
John Wiley and Sons, 1988

COMMENTS:

This book is fairly well written and easy to understand. However, some of the graphics leave a little to be desired. The early chapters are standard fare for this type of introductory text. The last third of the book discusses knowledge representation and does a good job of presenting several paradigms for representing and reasoning with knowledge. The rules chapter is terse, but the five chapters after it provide good coverage of their topics.

LEVEL: Introductory / Intermediate

INTENDED AUDIENCE: General

CONTENTS:

Getting Started
Forming the Development Team
Acquiring Knowledge
Developing the Requirements
Conducting the Feasibility Study
Crafting a System Design
Crafting a Prototype Application
Evaluating the Prototypes
Crafting the Pilot Application
Conducting the Operational Phases
Preparing a Project Schedule
Crafting Successful Knowledge-Based Systems
Knowledge Crafting with Rule-Based Representations
Knowledge Crafting with Frame-Based Reasoning
Knowledge Crafting with Multiple Contexts
Knowledge Crafting with Model-Based Representations
Knowledge Crafting with Blackboard Representations
Selecting Appropriate Knowledge Representation Techniques

Expert Systems and Fuzzy Systems
C. V. Negoita
Benjamin/Cummings, 1985

COMMENTS:

This text provides a good introduction to the concepts of approximate reasoning and fuzzy sets. While not providing the required breadth for an introductory course text, it is a very good reference for the portions of the course covering the subject of inexact reasoning. The text provides numerous examples to illustrate the concepts.

LEVEL: Introductory

INTENDED AUDIENCE: General

CONTENTS:

Introduction
Exact and Inexact Reasoning in Knowledge Engineering
Fuzzy Sets
Knowledge Representation
Approximate Reasoning
Knowledge Engineering in Decision Support Systems
Knowledge Engineering in Management Support Systems

Expert Systems - Principles and Case Studies
R. Forsyth, ed.
Chapman and Hall, 1985

COMMENTS:

This aim of this book is, as the author states, to explain "the concepts behind expert systems to computer users unfamiliar with the latest research." The text is clearly written and illustrated, but is lacking in breadth. This book has much more coverage of the PROLOG side of knowledge engineering than most introductory expert system texts. However, other forms of knowledge representation, such as frames, are not covered. Although interesting, the description of how an expert system shell was built is not particularly useful for an introductory text. This text would be best used as an additional course reference on how PROLOG is being used for expert system development. The text also contains a well-written and clear introduction to machine learning.

LEVEL: Introductory

INTENDED AUDIENCE: General

CONTENTS:

Background
    The Expert Systems Phenomenon
    The Architecture of Expert Systems
    The Fifth Generation Game
    How Shall We Judge an Expert System?
Inference
    Fuzzy Reasoning Systems
    How to Build an Inference Engine
Knowledge Engineering
    Knowledge Engineering in PROLOG
    How we Built Micro Expert
    REVEAL: An Expert Systems Support Environment
Learning
    Machine Learning Strategies
    Adaptive Learning Strategies
    Automating Knowledge Acquisition
    The Knowledge Industry

Expert Systems – Artificial Intelligence in Business
P. Harmon and D. King
John Wiley and Sons, 1985

COMMENTS:

The text is introductory and intended for a business audience, but a lot of the information presented is useful to a general audience. The presentation is extremely clear and understandable, but like most of these general audience texts, somewhat superficial. In addition, the coverage is primarily on the rule-based aspects of expert systems and barely touches other knowledge representations. The text does provide useful descriptions of the current commercially available knowledge-based system development tools for many different platforms.

LEVEL: Introductory

INTENDED AUDIENCE: Business / General

CONTENTS:

Introduction
MYCIN
Human Problem Solving
Representing Knowledge
Drawing Inferences
MYCIN Revisited
Languages and Tools
Commercial Tools
Early Systems
Recent Systems
Building a Small Knowledge System
How Large Knowledge Systems are
    Developed
The Market for Knowledge Engineering
Knowledge Systems in the Next Five Years
Expert Systems for Training
Preparing for the Knowledge Systems
    Revolution

Expert Systems for Civil Engineers: Technology and Application
M. L. Maher, ed.,
American Society of Civil Engineers, 1987

COMMENTS:

This book starts out with a brief description of expert system technology and then describes the various application of knowledge-based system technology to several sub-areas of civil engineering. Each application description consists of a general description of the problem domain, a brief description of the knowledge representation and inference methodologies applied to the problem, and the current status of the project. The text provides a fairly exhaustive list of the systems that have been developed within civil engineering and would thus serve well as a reference to knowledge-based system courses taught with an engineering emphasis.

LEVEL: Introductory / Intermediate

INTENDED AUDIENCE: Engineering / General

CONTENTS:

Expert System Components
Languages and Tools for Building Expert
   Systems
Implementation Issues in the Building of
   Expert Systems
Expert Systems in Structural Engineering
Expert Systems in Geotechnical
   Engineering
Expert Systems in Construction
Expert Systems in Environmental
   Engineering
Expert Systems in Transportation
   Engineering

Introduction To Expert Systems
P. Jackson
Addison-Wesley, 1986

COMMENTS:

This text goes into much greater detail about aspects of knowledge representation and inference than is normally encountered in introductory texts. The text provides an interesting case study on CENTAUR, which uses rules and frames to provide a hybrid knowledge representation strategy.

LEVEL: Intermediate

INTENDED AUDIENCE: Computer Science / General

CONTENTS:

Expert Systems and Artificial Intelligence
Heuristic Search in DENDRAL and
   META-DENDRAL
Production Systems
Structured Objects
Predicate Logic
MYCIN: Medical Diagnosis using Production
   Rules
MYCIN Derivatives: EMYCIN, TEIRESIAS
   and NEOMYCIN
INTERNEST: Representation of Knowledge
   by Structured Objects
R1: Recognition as a Problem Solving
   Strategy
CENTAUR: A Combination of Rules and
   Frames
Meta-level Inference and Common Sense Reasoning in MECHO
Tools for Building Expert Systems
Knowledge Acquisition
Explaining Expert System Behavior

Knowledge-based Systems In Artificial Intelligence
R. Davis and D. Lenat
McGraw-Hill, 1982

COMMENTS:

The book is composed of two research reports on projects done with the Stanford Heuristic Programming Project, each report comprising one of two parts. The first part, by Lenat, describes the representation and use of heuristics in AM – a system for developing new interesting mathematical concepts. The second part, by Davis, describes Teiresias – a system that explained and helped debug the knowledge within MYCIN by using meta-level knowledge. Both authors provide plenty of detailed, real examples of their knowledge representation strategies. This book is intended for a more experienced audience, with some background in artificial intelligence and knowledge representation. Because of its in depth coverage of both frame-based and rule-based representation and inference strategies in the context of two complex domains, the book would be effective for a higher-level course text or as a reference in an introductory course.

LEVEL: Intermediate

INTENDED AUDIENCE: Computer Science / General

CONTENTS:

Part 1 – AM: Discovery in Mathematics as
   Heuristic Search
      Overview
      Example: Discovering Prime Numbers
      Agenda
      Heuristics
      Concepts
      Results
      Evaluating AM
Part 2 – Teiresias: Applications of
   Meta-Level Knowledge
      Introduction
      Background

Explanation
Knowledge Acquisition: Overview
Knowledge Acquisition I
Knowledge Acquisition II
Strategies
Conclusions

Programming Expert Systems In OPS5
L. Brownston, R. Farrell, E. Kant and N. Martin
Addison-Wesley, 1985

COMMENTS:

This advanced text is aimed at experienced rule-based programmers in industries and universities. It does cover some general information information about rule-based expert systems, but the primary purpose of this text is, as the authors put it, "to provide an explanation of production system languages and programming styles." As the title indicates, the book discusses rule-based programming in the context of OPS5. It provides excellent discussions of strategies and techniques for programming rule-based expert systems and provides numerous examples to illustrate these concepts.

LEVEL: Advanced

INTENDED AUDIENCE: Rule-based Programmers

CONTENTS:

Computing with Production Systems
Programming Practice in OPS5
An Example in OPS5
Organization and Control in OPS5
Advanced Programming Techniques in OPS5
Efficiency in OPS5
Production System Architecture
Knowledge Acquisition, Learning and Explanation in Production Systems
Related Expert System Tools

The Guide To Expert Systems
A. Goodall
Learned Information, 1985

COMMENTS:

Most of the discussions in this text, although informative, are fairly shallow. It discusses several historical and well-publicized systems such as MYCIN, PROSPECTOR, DENDRAL, XCON, and MOLGEN, but does not go into detail about either the domain or the implementation of the systems, except for MYCIN.

LEVEL: Introductory

INTENDED AUDIENCE: General

CONTENTS:

What are Expert Systems?
Expert System Applications
The Reasoning Power of Expert Systems
The Advantages of Expert Systems
Inside Expert Systems
Building an Expert System
Who Builds Expert Systems?
The History of Expert Systems
The Future of Expert Systems
A Selective List of Companies Supplying Expert Systems Products and Services

3 ARTIFICIAL INTELLIGENCE TEXTS AND REFERENCES

The following section describes several textbooks and references on artificial intelligence. The descriptions are in the same format as those for knowledge- based expert systems texts.

Artificial Intelligence
E. B. Hunt
Academic Press, 1975

COMMENTS:

This text provides an in-depth coverage of the "mathematical and computational approaches to the problems in the artificial intelligence field." The emphasis of this text is more on the fundamental principles than on implementation details. It is a much more rigorous, mathematical coverage of the subject than is usually presented.

LEVEL: Intermediate

INTENDED AUDIENCE: Computer Science

CONTENTS:

Introduction
    The Scope of Artificial Intelligence
    Programming, Program Structure, and Computability
Pattern Recognition
    General Considerations in Pattern Recognition
    Pattern Classification and Recognition Methods Based on Euclidean Description Spaces
    Non-Euclidean Parallel Procedures: The Perceptron
    Sequential Pattern Recognition
    Grammatical Pattern Classification
    Feature Extraction
Theorem Proving and Problem Solving
    Computer Manipulable Representations in Problem Solving
    Graphic Representations in Problem Solving
    Heuristic Problem Solving Programs
    Theorem Proving
Comprehension
    Computer Perception
    Question Answering
    Comprehension and Natural Language
Review and Prospectus

Artificial Intelligence
E. Rich
McGraw-Hill, 1983

COMMENTS:

This text provides an excellent, clearly written introduction to the concepts and strategies of intelligent problem solving and knowledge representation. Sufficient detail is given to the description of problem solving methods such as forward and backward reasoning, matching, heuristic search, hill climbing, generate-and-test, and constraint satisfaction.

LEVEL: Introductory

INTENDED AUDIENCE: Computer Science / General

CONTENTS:

What is Artificial Intelligence?
Part 1 – Problem Solving
    Problems and Problem Spaces
    Basic Problem Solving Methods
    Game Playing
Part 2 – Knowledge Representation
    Knowledge Representation using Predicate Logic
    Knowledge Representation using Other Logics
    Structured Representations of Knowledge
Part 3 – Advanced Topics
    Advanced Problem Solving Systems
    Natural Language Understanding
    Perception
    Learning
    Implementing A.I. Systems: Languages and Machines
Conclusion

Artificial Intelligence, 2nd Edition
P. Winston
Addison-Wesley, 1985

COMMENTS:

This book provides a clear, modular coverage of the fundamentals of artificial intelligence. The algorithms are all described in English, allowing people not familiar with LISP to read and understand them, too. The book is extremely well written and well illustrated. The information and some of the examples have been updated to include recent advancements and application domains.

LEVEL: Introductory

INTENDED AUDIENCE: Computer Science / General

CONTENTS:

The Intelligent Computer
Description Matching and Goal Reduction
Exploiting Natural Constraints
Exploring Alternatives
Problem Solving Paradigms
Logic and Theorem Proving

Representing Commonsense Knowledge
Language Understanding
Image Understanding
Learning Class Descriptions from Samples
Learning Rules from Experience

Artificial Intelligence – An Introductory Course
A. Bundy
Edinburgh University Press, 1980

COMMENTS:

This text was one of the first on the subject of artificial intelligence and was used at Edinburgh University in their introductory artificial intelligence classes. In spite of its age, the text does cover many of the fundamental concepts of AI. However, its coverage of problem solving methods is not very broad and does not get much past means–ends analysis; production systems are barely mentioned. It is very well illustrated and provides numerous examples.

LEVEL: Introductory

INTENDED AUDIENCE: Computer Science / General

CONTENTS:

What is Artificial Intelligence?
Problem Solving
Natural Language
Question Answering and Inference
Visual Perception
Learning
Programming

Handbook of Artificial Intelligence, Vol. 1
A. Barr and E. Feigenbaum
Kaufmann, 1981

COMMENTS:

This three volume set is by far the most comprehensive reference to the various fields of artificial intelligence. The subject matter is presented so that the non-computer science professional can comprehend and use the handbook information. Each subject is well documented, making these books a very good starting point for literature reviews.

LEVEL: Introductory

INTENDED AUDIENCE: Computer Science / General

CONTENTS:

Introduction
Search
Knowledge Representation
Understanding Natural Language
Understanding Spoken Language

Handbook of Artificial Intelligence, Vol. 2
A. Barr and E. Feigenbaum
Kaufmann, 1982

COMMENTS: same as Vol. 1

LEVEL: Introductory

INTENDED AUDIENCE: Computer Science / General

CONTENTS:

Programming Languages for AI Research
Applications-oriented AI Research: Science
Applications-oriented AI Research: Medicine
Applications-oriented AI Research: Education
Automatic Programming

Handbook of Artificial Intelligence, Vol. 3
P. Cohen and E. Feigenbaum
Kaufmann, 1982

COMMENTS: same as Vol. 1

LEVEL: Introductory

INTENDED AUDIENCE: Computer Science / General

CONTENTS:

Models of Cognition
Automatic Deduction
Vision
Learning and Inductive Inference
Planning and Problem Solving

Introduction to Artificial Intelligence

E. Charniak and D. McDermott
Addison-Wesley, 1985

COMMENTS:

This text also explains many of the concepts of artificial intelligence in a very clear and understandable way. It has also been used fairly extensively in universities as a course text in introductory artificial intelligence courses. While the text does cover a wide range of topics, it still goes into sufficient depth of coverage by discussing, and providing psuedocode for, several algorithms for search and natural language processing.

LEVEL:	Introductory
INTENDED AUDIENCE:	Computer Science / General

CONTENTS:

AI and Internal Representation
LISP
Vision
Parsing Language
Search
Logic and Deduction
Memory Organization and Deduction
Abduction, Uncertainty and Expert Systems
Managing Plans of Action
Language Comprehension
Learning

Principles of Artificial Intelligence

N. Nillson
Tioga, 1980

COMMENTS:

This text is very well written, but does not provide the breadth of coverage found in more recent introductory artificial intelligence textbooks. It does, however, provide very in-depth descriptions of production systems, deduction strategies, and planning systems. This text would be appropriate for an introductory course in artificial intelligence, especially if the course is to have an emphasis on production systems, and would be ideal as a primary reference for an expert systems course.

LEVEL:	Introductory
INTENDED AUDIENCE:	Computer Science / General

CONTENTS:

Production Systems and AI
Search Strategies for AI Production Systems
Search Strategies for Decomposable Production Systems
The Predicate Calculus in AI
Resolution Refutation Systems
Rule-Based Deduction Systems
Basic Plan-Generating Systems
Advanced Plan-Generating Systems
Structured Object Representations

4 BIBLIOGRAPHY OF TEXTS AND REFERENCES

The following is an incomplete list of texts and references that address some aspect of knowledge-based systems or artificial intelligence. The list includes books addressing general theory as well as those addressing application of that theory. In this section, the texts are grouped by subject, and alphabetized within that grouped according to first author. The subjects by which the texts are grouped are: knowledge-based expert systems texts, artificial intelligence texts, and LISP and PROLOG programming.

This list was compiled from lists built by Gero (1985), Mantock (1986) and Waterman (1986), and recent publication lists from publishers. The listings ending with a "□" were briefly described in the previous sections.

4.1 Knowledge-Based Expert Systems: General

Brownston, L., Farrell, R., Kant, E. and Martin, N. (1985). *Programming Expert Systems in OPS5*. Addison-Wesley, Reading. □

Davis, R., and Lenat, D. B. (1982). *Knowledge-Based Systems in Artificial Intelligence*, McGraw-Hill, New York. □

Forsyth, R., ed. (1985). *Expert Systems Principles and Case Studies*, Chapman and Hall, London. □

Goodall, A. (1985). *The Guide to Expert Systems*, Learned Information, Oxford. □

Harmon, P., and King D. (1985). *Expert Systems – Artificial Intelligence in Business*, John Wiley and Sons, New York. □

Hayes-Roth, F., Waterman, D. A., and Lenat, D. B., eds. (1983). *Building Expert Systems*, Addison-Wesley, Reading. □

Jackson, P. (1986). *Introduction to Expert Systems*, Addison-Wesley, Reading. □

Martin, J., and Oxman, S. W. (1988). *Building Expert Systems – A Tutorial*, Prentice-Hall, Englewood Cliffs, New Jersey. □

Negoita, C. (1985). *Expert Systems and Fuzzy Systems*, Benjamin/Cummings, Menlo Park. □

Walters, J. R., and Nielson, N. R. (1988). *Crafting Knowledge-Based Systems – Expert Systems Made Realistic*, John Wiley and Sons, New York. □

Waterman, D. A. (1986). *A Guide to Expert Systems*, Addison-Wesley, Reading. □

Weiss, S., and Kulikowski, C. (1984). *A Practical Guide to Designing Expert Systems*, Rowen and Allenheld, Totowa, New Jersey. □

Williamson, M. (1986). *Artficial Intelligence for Micro-computers – The Guide for Business Decisionmakers*, Brady, New York. □

4.2 Knowledge-Based Expert Systems: Reference

Adeli, H., ed. (1987). *Expert Systems in Construction and Structural Engineering*, Chapman and Hall, New York.

Buchanan, B. G., and Shortliffe, E. H. (1985). *Rule-Based Expert Systems – The MYCIN Experiments of the Stanford Heuristics Programming Project*, Addison-Wesley, Reading. □

Clancy, W. J., and Shortliffe, E. H., eds. (1984). *Readings in Medical Artificial Intelligence: The First Decade*, Addison-Wesley, Reading.

Gero, J., ed. (1985). *Knowledge-Engineering in Computer-Aided Design*, North Holland, Amsterdam.

Gero, J., ed. (1987). *Expert Systems in Computer-Aided Design*, North Holland, Amsterdam.

Harmon, P, Maus, R., and Morrissey, W. (1988). *Expert Systems Tools and Applications*, John Wiley and Sons, New York.

Hart, A. (1986). Knowledge Acquisition for Expert Systems, McGraw-Hill, New York.

Kostem, C., Maher, M. L., eds. (1986). *Expert Systems for Civil Engineers*, American Society of Civil Engineers, New York.

Kowalik, J. S. and Kitzmiller, C. T. eds. (1988). *Coupling Symbolic and Numeric Computing in Expert Systems, II*, North Holland, Amsterdam.

Maher, M. L., ed. (1987). *Expert Systems for Civil Engineers: Technology and Application*, American Society of Civil Engineers, New York. □

McGraw, K. L., and Harbison-Briggs, K. (1989). *Knowledge Acquisition Principles and Guidelines*, Prentice-Hall, Englewood Cliffs, New Jersey.

Michie, D., ed. (1982). *Introductory Readings in Expert Systems*, Gordon and Breach, New York.

Michie, D., ed. (1979). *Expert Systems in the Micro-Electronic Age*, Edinburgh University Press, Edinberg.

Pham, D. T., ed. (1988). *Expert Systems in Engineering*, IFS, Springer-Verlag, Berlin.

Ralston, D. W. (1988). *Principles of Artificial Intelligence and Expert Systems Development*, McGraw Hill, New York.

Rychener, M. D., ed. (1988). Expert Systems for Engineering Design, Academic Press, New York.

Rauch-Hindin, W. B. (1985). *Artificial Intelligence in Business, Science, and Industry: Volume II - Applications*, Prentice-Hall, Englewood Cliffs, New Jersey.

Shortliffe, E. H. (1976). *Computer-Based Medical Consultations: MYCIN*, Elsevier, New York.

4.3 Artificial Intelligence: General

Bundy, A. (1980). *Artificial Intelligence - An Introductory Course*, revised edition, Edinburgh University Press, Edinburgh.

Charniak, E., and McDermott, D. (1985). *The Introduction to Artificial Intelligence*, Addison-Wesley, Reading.

Hunt, E. (1975). *Artificial Intelligence*, Academic Press, New York.

Mishkoff, H. C. (1985). Understanding Artificial Intelligence, Howard W. Sams and Co., Indianapolis.

Nillson, N. (1980). *Principles of Artificial Intelligence*, Tioga, Palo Alto.

Rich, E. (1983). *Artificial Intelligence*, McGraw-Hill Book Company, Artificial Intelligence Series, New York.

Winston, P. (1985). *Artificial Intelligence*, 2nd Ed., Addison-Wesley, Reading.

4.4 Artificial Intelligence: Reference

Barr, A., and Feigenbaum, E., eds. (1981). *Handbook of Artificial Intelligence*, Vol. 1, Kaufmann, Los Altos.

Barr, A., and Feigenbaum, E., eds. (1982). *Handbook of Artificial Intelligence*, Vol. 2, Kaufmann, Los Altos.

Bobrow, D. G., and Collins, A., eds. (1975). *Representation and Understanding - Studies in Cognitive Science*, Academic Press, Inc., Orlando.

Cohen, P., and Feigenbaum, E., eds. (1982). *Handbook of Artificial Intelligence*, Vol. 3, Kaufmann, Los Altos.

Feigenbaum, E. and Feldman, J., eds. (1963). *Computers and Thought*, McGraw-Hill, New York.

Feigenbaum, E., and McCorduck, P. (1983). *The Fifth Generation*, Addison-Wesley, Reading.

Forsyth, R. (1989). *Machine Learning, Principles and Techniques*, Chapman and Hall, London.

Forsyth, R., and Rada, R. (1986). *Machine Learning: Applications in Expert Systems and Information Retrieval*, John Wiley and Sons, Halsted Press Division, Ney York.

Grimson, W. E. L., and Patil, R. S., eds. (1987). *AI in the 1980s and Beyond*, MIT Press, Cambridge.

Latombe, J. C., ed. (1979). *Artificial Intelligence and Pattern Recognition in Computer-Aided Design*, North-Holland, Amsterdam.

McCorduck, P. (1979). *Machines Who Think*, Freeman, San Francisco.

Newell, A., and Simon, H. (1972). *Human Problem Solving*, Prentice-Hall, Englewood Cliffs, New Jersey.

Nillson, N. (1971). *Problem Solving Methods in Artificial Intelligence*, McGraw-Hill, New York.

Schank, R. C., and Abelson, R. P. (1977). *Scripts, plans, goals, and understanding*, Lawrence Erlbaum, Hillsdale, NJ.

Schank, R. C. (1982). *Dynamic Memory*, Cambridge University Press, Cambridge.

Schrobe, H. and The American Association of Artificial Intelligence, eds. (1988). *Exploring Artificial Intelligence*, Morgan Kaufmann, San Mateo.

Simon, H. (1981). *The Sciences of the Artificial*, 2nd Edn., MIT Press, Cambridge.

Taylor, W. A. (1988). *What Every Engineer Should Know About Artificial Intelligence*, MIT Press, Cambridge.

Winston, P., and Prendergast, K. (1984). *The AI Business*, MIT Press, Cambridge.

4.5 LISP and PROLOG

Abelson, H., and Sussman, G. J. (1985). *Structure and Interpretation of Computer Programs*, MIT Press, Cambridge.

Charniak, E., Riesbeck, C., and McDermott, D. (1979). *Artificial Intelligence Programming*, Erlbaum, Hillsdale, New Jersey.

Clocksin, W. F., and Mellish, C. S. (1981). *Programming in PROLOG*, Springer-Verlag, Berlin.

Steele, G. (1984). *Common LISP: The Language*, Digital Press, Burlington.

Sterling, L., and Shapiro, E. (1986). *The Art of PROLOG*, MIT Press, Cambridge.

Touretzky, D. (1984). *LISP: A Gentle Introduction to Symbolic Computation*, Harper & Row, New York.

Wilensky, R. (1984). *LISPCraft*, W. W. Norton, New York.

Winston P. H., and Horn, B. (1984). *LISP*, 2nd Ed., Addison-Wesley, Reading.

5 REFERENCES

Gero, J. S. (1985). "Bibliography of Books on Artificial Intelligence with Particular Reference to Expert Systems and Knowledge Engineering", *Computer-Aided Design*, 17(9).

Mantock, J. (1986). "An Annotated Bibliography for AI", *TI Engineering Journal*, January-February.

Chapter 13
Knowledge-Based Expert Systems In Civil Engineering: An Annotated Bibliography

by D. Sriram, A. M. ASCE

1 INTRODUCTION

The number of papers published on the applications of knowledge-based expert systems (KBES) to civil engineering problems in the last decade reflects the interest being shown in the community. The intent of this chapter is to provide an annotated bibliography of the applications of KBES in civil engineering. Articles, technical reports, and theses are covered in the next section. This is followed by a list of books and collections of papers.

The bibliography is by no means complete, and the author would appreciate pointers to other literature in the area for inclusion in a future update. The source for most of the doctoral thesis references is the *AI Related Dissertations* section, authored by Humphrey and Krovetz, of the SIGART newsletter.

2 ARTICLES

Adeli, H. and Balasubramanyam, K., A Knowledge-Based System for Design of Bridge Trusses, *ASCE Journal of Computing in Civil Engineering*, Vol. 2, No. 1, Pages 1-20, Jan. 1988 [See also Winter 1988 issue of the AI magazine].

> Describes BTEXPERT, which consists of an optimization program interfaced with an expert system, does optimal design of bridges. It incorporates both numerical and symbolic computing in a single framework.

Adeli, H., Knowledge-Based Expert Systems in Structural Engineering, In *CIVIL-COMP 85, Proceedings of the Second International Conference on Civil and Structural Engineering Computing*, London, England, pages 391-398, Dec. 3-5, 1985 [See also *Engineering Analysis*, Vol. 3, No. 3, pages 154-160, September 1986].

Ahmad, K. et al., Implementation Issues of Hydrological Expert Systems - A Civil Engineering Case Study, In *CIVIL-COMP 85. Proceedings of the Second International Conference on Civil and Structural Engineering Computing*, London, England, pages 407-414, Dec. 3-5, 1985.

Akiner, V.T., Knowledge-Based Systems for Tall Buildings, *Second Century of the Skyscraper*, Van Nostrand Reinhold Company, pages 591-604, 1988.

> Describes the basic problems of expert systems development as related to tall buildings, identifies main sources of knowledge to be used in the development of such systems, and gives examples of decision rules extracted from the monograph on tall buildings. A conceptual framework for a knowledge-based system for tall buildings is also presented.

Alkass, S. and Harris, F., Expert System for Earthmoving Equipment Selection in Road Construction, *ASCE Journal of Construction Engineering and Management*, Vol. 114, No.3, pages 426-440, Sept. 1988.

Alshawi, M. and Cabrera, An Expert System to Evaluate Concrete Pavements, *Microcomputers in Civil Engineering*, Vol. 3, No. 3, pages 191-197, 1988.

> Describes "Pavement Expert", which aids in the evaluation of concrete pavements. Pavement Expert is implemented on an IBM-PC in the SAVIOR shell.

Arciszewski, T., Mustafa, M., and Ziarko, W., A Methodology of Design Knowledge Acquisition for Use in Learning Expert Systems, *International Journal of Man-Machine Studies*, No. 37, pages 23-31, 1987.

> Provides an introduction to computer learning from examples based on the theory of rough sets. A methodology of inductive learning in proposed. Include a description of an inductive learning experiment conducted in the area of structural engineering. Inductive learning is discussed in the context of learning expert systems for structural design.

Arciszewski, T., Ziarko, W., Adaptive Expert Systems for Preliminary Design of Wind Bracings, *Second Century of Skyscraper*, Van Nostrand Reinhold Company, pages 847-855, 1988.

> Describes a concept of an adaptive expert system for the conceptual design of wind bracings in the

steel skeleton structures of tall buildings. A solution generator in the system is based on the stochastic simulation of morphological analysis, and the learning component uses inductive learning based on the theory of rough sets. Experiments on the generation of wind bracing concepts and their feasibility analysis are also described.

Balachandran, M. and Gero, J., A Model for Knowledge-Based Graphical Interfaces, In Gero, J. and Stanton, R. (editors), *Artificial Intelligence Developments and Applications*, Elsevier Science Publishers, pages 147-163, 1988.

> Describes a model for a knowledge-based graphical interface which incorporates domain knowledge. A prototype system, implemented in PROLOG and C on a SUN workstation, is described (a different approach to integration between CAD systems and KBES is discussed in the paper by Jain and Maher).

Bennett, J. and Engelmore R., SACON: A Knowledge-based Consultant for Structural Analysis, In *Proceedings Sixth IJCAI*, Morgan Kaufmann Publishers, Inc., pages 47-49, 1979 [For a more detailed description see Technical Report STAN-CS-78-699, Stanford University, September 1978].

> SACON is an expert program that advises a structural engineer in the use of modeling options for MARC, a non-linear structural analysis program. It is implemented in EMYCIN. It does not have any interface with the analysis program.

Bonnet, A. and Dahan, C., Oil-Well Data Interpretation Using Expert System and Pattern Recognition Techniques, In *Proceedings Eighth IJCAI*, pages 185 - 189, 1983.

> Describes LITHO, which is a program for interpreting oil well data. The output from the program is a litholog, a description of rocks encountered in a well. LITHO is being developed at Schlumberger, France. The knowledge-base contains about 500 rules.

Bremdal, B.A., Control Issues in Knowledge-Based Planning System for Ocean Engineering Tasks, *Proceedings of 3rd International Expert Systems Conference*, London, June, 1987, pages 21-36. [See also Bremdal, B.A., *A Knowledge-Based Proposer for Offshore Installation Works. The Mark I version,* Technical Report No. MP/R 3, Dept. of Marine Technology, The Norwegian Institute of Technology, The University of Trondheim, 1987].

> LIFT-2 generates plans for heavy-lift operations. LIFT-2 generates the schedule using a hierarchy of goals to guide the process. It uses constraint satisfaction, coupled with the generate and test paradigm to find ways to achieve goals and subgoals. At lower levels of the goal hierarchy, LIFT-2 resorts to the *least commitment principle*. In effect, the program is able to postpone decisions until sufficient information is available. A recent version of the program - LIFT-3, which is implemented as a blackboard system - plans jobs with multiple contractors.

Chehayeb, F., *A Framework for Engineering Knowledge Representation and Problem Solving*, Ph.D. Thesis, Dept. of Civil Engineering, M.I.T., 1987 [See also Microcomputers in Civil Engineering, Vol. 4, No. 1, March 1989].

> An innovative structural design system - SIDS - implemented in GEPSE, a knowledge-based framework for engineering applications, is described. SIDS achieves innovative design concepts by combining existing components in a novel manner.

Chan, W. T. and Paulson, N. C. Jr., *Exploratory Design Using Constraints*, Journal of Artificial Intelligence in Engineering, Manufacturing, and Design, Vol. 1, No. 1, pages 59-71, 1988.

> A constraint management facility for design is developed in PROLOG.

Chen, J. L. and Hajela, P., FEMOD: A Consultative Expert System for Finite Element Modeling, *Computers and Structures*, Vol. 29, No. 1, pages 99-109, 1988 [See also the paper by Turkiyyah and Fenves].

> Describes FEMOD, which is an intelligent front to a special purpose finite element program - EAL. FEMOD contains rules for selection of elements, node selection, mesh generation, selection of subdivision lines, and mesh refinement.

Chiou, W. C., NASA Image-Based Geological System Development Project for Hyperspectral Image Analysis, *Applied Optics*, Vol. 24, No. 14, pages 2085-2091, July 1985.

Cobb, J. E., *A Microcomputer Approach to Contract Management Using AI*, Unpublished Master's Thesis, University of Colorado, Boulder, CO, 1984.

> The knowledge-base and logic for the development of DSCAS, which is intended to provide legal advice for construction claims, is developed. Currently DSCAS is designed for "Differing Site Conditions" clause of the U. S. Government standard general conditions to a construction contract.

Cohn, L. F., Harris, R. and Bowlby, W., Knowledge Acquisition for Domain Experts, *ASCE Journal of Computing in Civil Engineering*,

Vol. 2, No. 2, April 1988.

> Knowledge acquisition techniques used in the development of CHINA, a KBES for highway noise barrier design, are presented.

Courteille, J. M., Fabre, M. and Hollander, C. R., An Advanced Solution: The Drilling Advisor, SECOFOR in *Proceedings 58th Annual Technical Conference and Exhibition, Society of Petroleum Engineers and AIME*, San Francisco, CA, October 1983.

Cuena, J., The Use of Simulation Models and Human Advice to Build an Expert System for the Defense and Control of River Floods, In *Proceedings Eighth IJCAI*, Morgan Kaufman, pages 246-249, 1983.

> A conceptual framework for an expert system to aid in the operation of flood control and plan civil defense in flood prone areas is provided. The rules are described based on a set of mathematical models. System currently pursued by Spanish Ministry of Public Works.

David, H., An Analysis of Expert Thinking, *International Journal Man-Machine Studies*, Vol. 18, pages 1-47, 1983.

> Deals with how human experts acquire, understand and use knowledge in the domain of geology, in particular petroleum geology.

De La Garza, J. and Ibbs, W., *A Knowledge Engineering Approach to the Analysis and Evaluation of Schedules for Mid-Rise Construction*, Civil Engineering Studies, Construction Research Series No. 23, Dept. of Civil Engineering, University of Illinois at Urbana-Champaign, July 1988.

> The formalization of scheduling knowledge for construction is addressed in this work. Implementation details of KBES in PC Plus and ART are also described.

Diekmann, J. E. and Kraiem, Z., *Explanation of Construction Engineering Knowledge in Expert Systems*, ASCE Journal of Construction Engineering and Management, Vol. 114, No.3, pages 364-389, Sept. 1988 [See also the paper in the April 1988 issue of ASCE Journal of Computing in Civil Engineering].

East, W., Knowledge-Based Approach to Project Scheduling Selection, *ASCE Journal of Computing in Civil Engineering*, Vol. 2, No. 4, pages 307-328, Oct. 1988.

> A KBES, developed at U.S. CERL, for selecting a scheduling system is presented.

Eastman, C. M., Automated Space Planning, *Artificial Intelligence*, Vol. 4, No. 1, Spring 1973 [For related papers see also *Spatial Synthesis in Computer-Aided Building Design*, edited by Eastman, 1975].

> One of the first papers that addresses the application of heuristics to space planning.

Fawcett, W., An Elementary Rule Interpreter for Architectural Design, In Smith, A. (editor), *Knowledge Engineering and Computer Modeling in CAD*, pages 259-269, Butterworths, London, 1986.

> Shape rules allow the generation of architectural designs; these are "if ... then ...' type rules where both the left and right hand sides are shapes. The system was installed within the microcomputer CAD system, AutoCAD.

Faghri, A., Joshua, S., and Demetsky, M., Expert System for the Evaluation of Rail/Highway Crossings, *ASCE Journal of Computing in Civil Engineering*, Vol. 2, No. 1, pages 21-37, Jan. 1988.

> A KBES for evaluating and prioritizing highway/railroad grade crossings for safety improvements in Virginia is described.

Fenves, et al., An Integrated Software Environment for Building Design and Construction, In *Proceedings of the Fifth Conference on Computing in Civil Engineering*, ASCE, March 1988.

> Describes IBDE, which integrates various phases of structural engineering, i.e., from architectural design to construction planning.

Fenves, S. J., A Framework for a Knowledge-Based Finite Element Assistant, In Dym, C. L. (editor), *Applications of Knowledge-Based Systems to Engineering Analysis and Design*, American Society of Mechanical Engineers, pages 1-8, 1985.

> Describes the features needed in a complete expert system to assist in the use of finite element analysis programs

Fenves, S.J., Maher, M. L. and Sriram, D., *Knowledge-Based Expert Systems in Civil Engineering*, IABSE Journal, J-29/85, pages 63-72, November, 1985 [See also ASCE Civil Engineering Magazine, October 1984].

> Explores the potential benefits of KBES in civil engineering.

Fjellheim, R. and Syversen, P., *An Expert System for SESAM-69 Program Selection*, Computas Report 83-6010, January 1983. (A. S. Computas, P. O. Box, 310, 1322 HOVIK, Norway)

> Describes an expert system front end for a large finite element program SESAM-69, developed by A. S. Computas. Patterned after SACON. Implemented in EMYCIN.

Forde, B., *An Applications of Selected Artificial Intelligence Techniques to Engineering Analysis*, Ph. D. thesis, Department of Civil Engineering, The University of British Columbia, Vancouver, V6T 1Y3, Canada, 1989 [See also Computers and Structures, Vol. 29, No. 1, pages 171-174, 1988].

> Describes an intelligent finite element analysis program called "SNAP". The primary difference between SNAP and conventional analysis programs is that SNAP employs an event-driven architecture that processes anaysis tasks in any order required by the problem being currently addressed. SNAP is implemented in an object oriented framework on a Macintosh.

Franck, B. and Krauthammer, T., Development of an Expert System for Preliminary Risk Assessment of Existing Concrete Dams, *Engineering With Computers*, Vol. 3, No. 3, pages 137-148, 1988 [See also 1989 issues of Engineering With Computers for detailed descriptions of the project].

> A PC Plus-based system that assists in the field inspection of existing concrete dams is described.

Furuta, H., Tu, K. and Yao, J. T. P., Structural Engineering Applications of Expert Systems, *Computer Aided Design*, Vol 17, No. 9, pages 410-419, Nov. 1985.

> Describes expert systems to assess safety during construction and to assess damage to a structure after earthquake.

Garrett, J. H., *SPEX: A Knowledge-Based Standards Processor for Structural Component Design*, Ph.D. Thesis, Dept. of Civil Engineering, Carnegie-Mellon University, Pittsburgh, PA 15213, 1986 [See also International Journal for AI in Engineering, Vol. 1, No. 1, July 1986].

> SPEX is a knowledge-based standard independent system that takes the parameters of a design, identifies the appropriate specifications and checks these designs for confirmation of these specifications. It is implemented in the Framekit/Rulekit environment developed at CMU and is based on the blackboard architecture.

Gaschnig, J., Reboh, R. and Reiter, J., *Development of a Knowledge-Based Expert System for Water Resource Problems*, Technical Report SRI Project 1619, SRI International, August 1981.

> Describes an intelligent interface (HYDRO) for selecting numerical values of parameters that are input to a simulation program (HSPF).

Gero, et al., *Knowledge Engineering Publications: Abstracts*, May 1988, Department of Architectural Science, University of Sydney, NSW 2006, AUSTRALIA.

> Abstracts for 90 publications are listed in this report.

Gero, J. S. and Radford, A. D., Knowledge Engineering in Computer Graphics, In *First Australian Conference on Computer Graphics*, Sydney, Australia, August 31-September 2, 1983.

Gero, J. S., Radford, A. D., Coyne, R., and Akiner, V. T., Knowledge Based Computer-Aided Architectural Design, In Gero, J. (editor), *Knowledge Engineering in Computer Aided Design*, pages 57-58, North-Holland, Elsevier Science Publishing Company, 1985.

> The role of knowledge based systems in architectural design is discussed.

Gero, J. S. and Coyne, R., The Place of Expert Systems in Architecture, In *Proceedings CADD-84*, Butterworth & Co., U. K., 1984.

> An introduction to expert systems, along with some prototype applications. Implications to synthesis are explored.

Goel, V. and Pirolli, P., Motivating the Notion of Generic Design within Information-Processing Theory: The Design Problem Space, *AI Magazine*, pages 18-38, Spring 1989.

> Presents a cognitive model of design problem solving.

Gosling, G., Application of Expert Systems in Air Traffic Control, *ASCE Journal of Transportation Engineering*, Vol. 113, No. 2, pages 139-154, March, 1987.

> Because of the complexity of the Air Traffic Control (ATC) system, expert systems may provide a valuable addition to human decision making and conventional data processing techniques. A number of potential applica- tions, including traffic flow management, controller support functions, system failure management, training, and system configuration planning, are described.

Gross, M., *Design as Exploring Constraints*, Ph.D. Thesis, Dept. of Architecture, M.I.T., 1986 [See also papers by Chan and Paulson, and Navinchandra].

>Proposes a theory of design that views design as the exploration of alternative sets of constraints and of the regions of alternative solutions they bound. These constraints represent design rules, relations, and objectives. A computational model supporting the theory is described, with implementation details.

Hadipriono, F. C. and Toh, H. S., Approximate Reasoning Models for Consequences on Structural Components Due to Failure Events, *Civil Engineering and Practical Design Engineering*, Vol. 5, No. 3, pages 155-170, 1986.

>Describes methods of dealing with qualitative descriptions of the reliability of parts used in structures using fuzzy set techniques.

Haren, P. and Montalban, M., Prototypical Objects for CAD Expert Systems, *IEEE COMPINT - Computer Aided Technologies*, pages 53-55, 1985.

>A blackboard system for harbor and breakwater design is described.

Harris, R. A., *Development of an Expert System to Control a Highway Noise Barrier Design Model*, Ph.D. Thesis, Vanderbilt University, 1985 [University Microfilms order no. ADG85-22423, See also ASCE Journal of Transportation Engineering, Vol. 113, No. 2, pages 127-138, 1987].

>Describes CHINA, which aides an engineer in the acoustical design of highway noise barrier. CHINA is closely coupled with a FORTRAN program (OPTIMA), developed for highway noise barrier design.

Haywood, G. G., A Rule-Based Program to Construct an Influence Line for a Statically-Determinate Multi-Span Beam Structure, In Adey, R. (editor), *Engineering Software IV*, Computational Mechanics Publications, Springer-Verlag, Berlin-Heidelberg, New York, pages 5-3 to 5-11, 1985.

Hendrickson, C., et al., *Expert System for Construction Planning*, ASCE Journal of Computing, Vol. 1, No. 4, pages 253-269, 1987.

>This paper describes CONSTRUCTION-PLANEX, which generates construction plans. The program suggests technologies, generates activities, determines precedences, estimates durations, and develops the schedule. CONSTRUCTION-PLANEX takes as input the specifications of the physical elements in the design, site information, and resource availability and produces as output a complete plan, a provisional schedule, and a cost estimate. The program's knowledge-base consists of a large number of knowledge sources. Each knowledge source is input in the form of a decision table, which is later converted into a network of frame schemas. The tabular input format allows for easy input and updating of the knowledge-base. A good user interface and an impressive graphical simulation of the project are also provided.

Hendrickson, C., Martinelli, D., and Rehak, D., Hierarchical Rule-based Activity Duration Estimation, *ASCE Journal of Computing in Civil Engineering*, Vol. 113, No. 2, pages 288-301, June 1987.

Holt, R. H., Adding Intelligence to Finite Element Modeling, In *Expert Systems in Government Symposium*, McLean, Virginia, pages 326-337, October 22-24, IEEE Press, 1986.

>This system, which can be used with multiple finite element codes, interfaces to a CAD/CAM model and helps recommend element types, model nature, and determines the axis of symmetry for beam elements.

Howard, C., *Integrating Knowledge-Based Systems with Database Management Systems for Structural Engineering Applications*, Ph.D. thesis, Dept. of Civil Engineering, Carnegie Mellon University, Pittsburgh, PA 15213, pages 138, September 1986 [See also special issue of Computer Aided Design, November, 1985 and Proceedings, AAAI-87].

>KADBASE is a knowledge-based framework for integrating heterogeneous CAE tools. It consists of a central interface processor and several KBES and DBMS. Syntactic and semantic mappings between various modules are achieved through KADBASE-provided mapping functions. The user can taylor these functions to his application.

Hutchinson, P. J., Rosenman, M. A., and Gero, J. S., RETWALL: An Expert System for the Selection and Preliminary Design of Earth Retaining Structures, *Knowledge-Based Systems*, Vol. 1, No. 1, pages 11-23, 1987.

>This paper describes an expert system for the selection and preliminary design of engineering earth retaining structures. it describes the domain and how the knowledge was acquired from textbooks, questionnaires and interviews. Details of the implementation of RETWALL using the expert system shell BUILD are provided as is a full script of a session.

Ishizuka, M., Fu, K. S. and Yao, J. T. P., Rule-based Damage Assessment System for Existing Structures, *SM Archives*, Vol. 8, pages 99-118, Martinus Nijhoff Publishers, The Hague, 1983 [See also the following technical reports from the Department of Civil Engineering, Purdue: CE-STR-81-5 and CE-STR-81-36].

> The above paper describes SPERIL-I, a rule-based expert system. SPERIL-I addresses the issue of damage assessment of structures after earthquakes and other possible hazardous events.

Jain, D. and Maher, M., Combing Expert Systems and CAD Techniques, In Gero, J. and Stanton, R. (editors),*Artificial Intelligence Developments and Applications*, Elsevier Science Publishers, pages 65-81, 1988 [See also Microcomputers in Civil Engineering, Vo. 3, No. 4, 1988].

> Discusses various issues involved in the integration of CAD packages, such as AutoCAD, with KBES. The graphical interfaces were developed in AutoCAD, while a KBES for evaluating structural designs was developed in EXSYS^{tm}.

Johnston, D.M., and R.N. Palmer, Application of Fuzzy Decision Making: An Evaluation, *Journal of Civil Engineering Systems*, Vol. 5, No. 2, June, pages 87-92, 1988.

> This paper addresses the modeling of subjective information in decision making. Two approaches commonly used in fuzzy sets are evaluated. The ability of the methods to replicate decision behavior is measured by the rank correlation between oderings of alternatives obtained by decision makers and those obtained by these methods.

Jones, M. and Saouma, V., Prototype Hybrid Expert System for R/C Design, *ASCE Journal of Computing in Civil Engineering*, vol. 2, No. 2, pages 136-143, April 1988.

> Teknowledge's M.1 was used to develop a rule-based system for structural design of reinforced concrete. This system was interfaced with algorithms coded in FORTRAN and C.

Jozwiak, S. F., Application of Artificial Intelligence Notions in Structural Optimization Programs, *Computers and Structures*, Volume 24, No. 6, pages 1009-1013, Pergamon Press, 1986.

> Describes a system that examines previous designs to assist in optimization. The system works with plane and space trusses.

Keirouz, W., *Domain Modeling of Constructed Facilities for Robotic Applications*, Ph. D. thesis, Department of Civil Engineering, Carnegie Mellon University, Pittsburgh, PA 15213, pages 161, September 1988.

> Object-oriented modeling of constructed facilities is explored. This type of modeling is useful for incorporating structure and behavior of physical systems. The utility of this model in robot planning is described.

Kelway, P. S., Architecturally Speaking: The Use of Speech Technology in Architectural Design, In Smith, A. (editor), *Knowledge Engineering and Computer Modeling in CAD*, pages 440-447, Butterworths, London, 1986.

> Discusses the integration of voice control into a computer-assisted architectural design system.

Kruppenbacher, T. A., *The Application of Artificial Intelligence To Contract Management*, Unpublished Master's Thesis, University of Colorado, Boulder, CO, 1984 [also US Army Corps of Engineers CERL, Technical Manuscript P-166, August 1984].

> Describes the implementation details of the Differing Site Condition Analysis System (DSCAS), which is implemented in ROSIE. See reference by Cobb.

Kuo, T. B., *Well Log Correlation Using Artificial Intelligence*, Ph. D. thesis, Texas A&M University, 1986. (University of Microfilms order no. ADG86-25409)

Lannuzel, P. and L. Ortolano, Evaluations of a Heuristic Program for Scheduling Treatment Plant Pumps, *Journal of Water Resources Planning and Management*, Vol. 115, July 1989.

> Describes PILOTE, a computer program for reproducing decisions of expert operators in scheduling outlet pumps at a water treatment plant near Paris, France. Emphasis is on testing the extent to which the program reproduces decisions of the experts on test case problems.

Lansdown, J., *Expert Systems: Their Impact on the Construction Industry*, RIBA Conference Fund, U. K., 1982.

> Presents a number of potential applications for KBES in the construction industry.

Levitt, R. E., Kartam, N. A., and Kunz, J. C., Artificial Intelligence Techniques for Generating Construction Project Plans, ASCE Journal of Construction Engineering and Management, Vol. 114, No.3, pages 329-343, Sept. 1988 [See also paper by Levitt and Kunz in AIEDAM, Vol. 1, No. 11, 1988].

> A good survey of AI applications in construction

planning.

Levitt, R. E. and Kunz, J. C., A Knowledge-Based System for Updating Engineering Project Schedules, In Dym, C. L. (editor), *Applications of Knowledge-Based Systems to Engineering Analysis and Design*, American Society of Mechanical Engineers, pages 47-65, 1985.

> Describes the PLATFORM project, developed at Stanford University. PLATFORM, which updates the schedule for the construction of offshore platforms, is implemented using the KEE^{tm} object-oriented language. The activities are represented as objects and are related through links; these links facilitate the updating of the project network. Rules are used to depict the project manager's expertise, such as assessing the affects of progress of a current activity on future activities. The use of PLATFORM is restricted to the processing of networks; it does not have the capability to generate the schedule for the project.

Li, H-L, An Expert System for Evaluating Multicriteria Transportation Networks, *Microcomputers in Civil Engineering*, Vol. 3, No. 3, pages 199-214, 1988.

> An hybrid KBES, developed using Lotus 123, Turbo PROLOG, and C on an IBM-PC, for evaluating transportation network improvement alternatives with multicriteria is presented.

Logcher, R., Sriram, D., and Cherneff, J., *Knowledge-Based Construction Scheduling*, Architectural & Engineering Systems, pages 18-20, February, 1989.

> Generating and maintaining schedules from architectural drawings is vital to both construction and design professionals. Automating the task requires combining a CAD system with knowledge based programming and database techniques. A prototype system, called BUILDER, is presented for the domain of routine interior office construction. The principal features of BUILDER are its knowledge representation scheme and its planning paradigm. A knowledge representation scheme is demonstrated that enables the interplay of the drawing and the construction knowledge base. The features of the drawing are extracted and represented in a semantic network of frames. A forward chaining rule-based mechanism reasons about the semantic network of frames to produce a project network. The project network is represented in an object oriented style that can readily admit functional extension. An object oriented CPM algorithm for scheduling the project network is also implemented.

Lopez, L. A., Elam, S. L. and Christopherson, T., SICAD: A Prototype Implementation System for CAD, In *Proceedings of the ASCE Third Conference on Computing in Civil Engineering*, San Diego, California, pages 84-93, April 1984.

> Describes a framework for the development of a KBES for providing an interface between standards governing engineering design and CAD programs.

Ludvigsen, P. and Dupont, R., Formal Evaluation of the Expert System DEMOTOX, *ASCE Journal of Computing in Civil Engineering*, Vol. 2, No. 4, pages 398-412, October 1988.

> DEMOTOX assesses groundwater contamination potential. This paper presents the evaluation and validation of DEMOTOX.

MacCallum, K. J., A Knowledge-base for Engineering Design Relationships, In *Expert Systems 82*, Technical Conference of the BCS SGES, U. K., 1982 [See also MacCallum, K. J., Creative Ship Design by Computer, In Rogers, D. F., Nehrling, B. C. and Kuo, C. (editors), *Computer Applications in the Automation of Shipyard Operation and Shipyard Design IV*, IFIP82, North-Holland Publishing Company, 1982].

> Deals with the development of a KBES for ship design. Also attempts to incorporate an element of learning in the system.

Mackenzie, C. A. and Gero, J. S., Learning Design Rules from Decisions and Performances, *Artificial Intelligence in Engineering*, Vol. 2, No. 1, pages 2-10, 1987.

> This paper examines an approach to the extraction of implicit knowledge in rule form about the relationships between design decisions and their performance consequences. The use of the methodology for learning about decision/performance relationships in extant designs is proposed.

Mackenzie, C. A., Inducing Relational Grammars from Design Interpretations, In Gero, J. and Stanton, R. (editors),*Artificial Intelligence Developments and Applications*, Elsevier Science Publishers, pages 315-328, 1988.

> Describes an experiment that combines the use of a heuristic driven search and a tree system inference technique to generate context-free design grammars.

Maher, M. L., Sriram, D. and Fenves, S. J., Tools and Techniques for Knowledge-based Expert Systems for Engineering Design, *Advances in Engineering Software*, Vol. 6, No. 4, Computational Mechanics Publications, October 1984.

> Describes the application of OPS5, SRL and PROLOG to engineering design problems.

Maher, M. L., and Fenves, S. J., HI-RISE: An Expert System For The Preliminary Structural Design Of High Rise Buildings, in *Knowledge Engineering in Computer-Aided Design*, Edited by J. Gero, North Holland, pages 125-146, 1985.

> HI-RISE is a KBES for preliminary structural design of high rise buildings. It uses a top-down refinement strategy, combined with a limited form of the constraint handling technique.

Maher, M., HI-RISE and Beyond: Directions for Expert Systems in Design, *Computer Aided Design*, Vol 17, No. 19, pages 420-426, November 1985.

> Describes extensions to HI-RISE.

Maher, M., Engineering Design Synthesis: A Domain Independent Representation, *Artificial Intelligence for Engineering, Design, and Manufacturing*, Vol. 1, No. 3, 1988.

> Describes a decomposition approach to synthesis and the implementation of EDESYN, a domain independent knowledge-based design programming environment.

Maher, M.L., Fenves, S.J., Rule of Expert Systems in High-Rise Building Design, *Second Century of the Skyscraper*, Van Nostrand Reinhold Company, pages 571-589, 1988.

> Basic concepts of expert systems are discussed. Hi-Rise, an expert system for the preliminary structural design of tall buildings, is described. Potential applications of expert systems in the area of tall buildings design are also discussed.

Manheim, M. L., *Hierarchical Structure: A Model of Design and Planning Processes*, MIT Press, Cambridge, Mass., 1966.

> The concept of hierarchical design was first implemented by Manheim for determining highway locations. A classical work in the area.

Markusz, Z., Design in Logic, *Computer Aided Design*, Vol. 14, No. 6, pages 335-343, November 1982. (other references to this work can be found in *Logic Programming*, Clark, K. L. and Tarnlund, S. A. (editors), Academic Press, 1982.)

> Describes the use of logic in architectural design. Implementation language is PROLOG.

Maser, K., Automated Interpretation for Sensing In-Situ Conditions, *ASCE Journal of Computing in Civil Engineering*, Vol. 2, No. 3, pages 215-238, July 1988.

> AISD is a program that uses knowledge-based techniques for interpretation of sensory data in the domain of pavement condition characterisation. Incoming signals are initially processed by signatures using algorithms and the interpretation is done by a blackboard-based KBES.

McLaughlin, S. and Gero, J. S., Acquiring Expert Knowledge from Characterised Designs, *Artificial Intelligence for Engineering Design, Analysis and Manufacturing*, Academic Press, Vol. 1, No. 2, pages 73-87, 1987.

> The induction algorithm ID3 is used as a means of inferring general statements about the nature of solutions which exhibit Pareto optimal performance in terms of a set of performance criteria. The rules inferred in a building design domain are compared with those extracted using a heuristic based learning system.

Medoff, S. M., Register, M. S., and Swartwout, M. W., Representing Knowledge for Design Verification and Evaluation Systems, In *AI'87 - Australian Joint AI Conference*, 1987.

> Describes CANDLE, a knowledge representation language for processing constraints. CANDLE was used to develop PREDICTE, which is a KBES for building estimation. CANDLE and PREDICTE are developed by Digital Equipment Corporation.

Melosh, R. J., Marcal, P. V. and Berke, L., Structural Analysis Consultation using Artificial Intelligence, In *Research in Computerized Structural Analysis and Synthesis*, NASA, Washington, D. C., October 1978.

> Illustrates an application of SACON.

Moore, G. R., An Applicable Model Theory, In Smith, A. (editor), *Knowledge Engineering and Computer Modeling in Cad*, pages 283-291, Butterworths, London, 1986.

> Describes an attempt to provide some theoretical basis to computer modeling.

Navinchandra, D., Sriram, D., and Logcher, R., GHOST: A Project Network Generator, *ASCE Journal of Computing in Civil Engineering*, Vol. 2, No. 3, pages 239-254, July 1988.

> GHOST takes as input a set of activities and produces as output a schedule by setting up precedences among the activities. The knowledge-base is made up of several knowledge sources known as *Critics*. The *Critics* contain knowledge about basic physics, construction norms, redundancy in networks, etc.. GHOST was implemented in IMST, a rule-based programming environment developed at M.I.T.

Navinchandra, D., *Exploring Innovative Designs*

by Relaxing Criteria and Reasoning from Precedent Knowledge, Ph.D. Thesis, Dept. of Civil Engineering, M.I.T., 1987.

> CYCLOPS (Criteria Yielding, Consistent Labeling Optimization and Precedents-based System) was developed to address the following issues: 1) designers work with goals, objects, and constraints; 2) designers are not mere satisficers; 3) new criteria emerge as design progresses; 4) past design examples are used extensively; and designers continue to innovate. CYCLOPS, whose domain is landscape design, operates in three modes (stages): *normal search*, *exploration*, *adaptation* stage; in the adaptation stage, it uses case-based reasoning techniques.

Navinchandra, D., *Intelligent Use of Constraints for Activity Scheduling*, CERL- N-86/15, Construction Engineering Research Laboratory, Illinois, July 1986.

> The use of constraints in scheduling is discussed.

Norabhoompipat, T. and Fenves, S. J., *An Information Processing Model of Civil Engineering Design Systems*, Technical Report R-78-110, Department of Civil Engineering, Carnegie-Mellon University, August, 1978 [See also the IFIPS proceedings edited by J-C-Latombe].

> Applications of artificial intelligence techniques in civil engineering are explored. Emphasis is placed on use of AI techniques for for specification representation and processing.

Oey, K. H., DARC: A Knowledge Based Design Assisting Representation Concept, In Smith, A. (editor), *Knowledge Engineering and Computer Modeling in CAD*, pages 270-281, Butterworths, London, 1986.

> Describes a frame-based system to assist in the early phases of architectural design.

Ohsuga, S., A New Method of Model Description - Use of Knowledge Base and Inference, In Bo, K. and Lillehagen, F. M. (editors), *CAD Systems Framework*, IFIP83, North-Holland Publishing Company, 1983.

> A methodology to represent the model building process in building design is proposed. Knowledge is represented in terms of expanded predicate logic and interfaced with a relational database.

Ortolano, L. and Perman, C.D., A Planners Introduction to Expert Systems, *Journal of the American Planning Association*, Vol. 53, No. 1, pages 98-103, Winter 1987.

> Introduces KBES and describes actual and potential applications in city and regional planning.

Oxman, R. and Gero, J. S., Using an Expert System for Design Diagnosis and Design Synthesis, *Expert Systems: The International Journal for Knowledge Engineering*, Vol. 4, No. 1, pages 4-15, 1987.

> This paper describes the concepts which allow an expert system to be used for both design diagnosis and design synthesis. An example of the implementation of these concepts is presented in the domain of preliminary design of domestic kitchens in the expert system PREDIKT.

Paek, Y., and Adeli, H., Representation of Structural Design Knowledge in a Symbolic Language, *ASCE Journal of Computing in Civil Engineering*, Vol. 2, No. 4, pages 346-364, Oct. 1988.

> Describes STEELEX, a KBES for detailed design of multistory rectangular steel buildings. STEELEX is implemented in SDL. SDL is a Structural Design Language developed in INTERLISP.

Palmer, R.N., Application of Artificial Intelligence to Civil Engineering, *Proceedings of ASCE Specialty Conference on Computer Applications in Water Resources*, Buffalo, NY, pages 591-600, June 1985.

> This paper provides a very general introduction to artificial intelligence and its application to civil engineering problems. Emphasis is placed on expert systems. Successful applications are noted with suggestions for potential areas of likely success.

Palmer, R. N., and Tull, R. M., Expert System for Drought Management Planning, *ASCE Journal of Computing in Civil Engineering*, ASCE, Vol. 1, No. 4, pages 284-297, October, 1987.

> This paper describes a prototype expert system designed to provide water supply managers with aid in developing water use restriction policies during periods of low flow. An expert system developed using Insight 2+ is used to allow managers to evaluate the degree to which existing drought conditions are similar to those that occurred in the past. The expert systems then aids the user in developing appropriate management policies based on the time of year, drought intensity and other system characteristics.

Palmer, R. N., Holmes, K. J., Menard, R., and Parkinson, D., An Expert System for Drought Management, *Critical Water Issues and Computer Applications, Proceedings of the 15th Annual Water Resources Conference*, ASCE , Norfolk, VA, pages 1-4, June 1988.

> This paper describes the development of SID, an expert system for drought management for the Seattle Water Department. Emphasis is placed on

the perspective of the water managers and the interviews required to develop the knowledge base of the expert system. The difficulties in developing such a knowledge base are noted with suggestions to make the process more efficient.

Palmer, R. N., and Mar, B. W. Expert System Software for Civil Engineering Applications, *Journal of Civil Engineering Systems*, Vol. 4, No. 4, 1988.

This paper examines the advantages that expert systems offer over more conventional programming, describes the basic components of such systems, and suggests areas of applications that are likely to be successful. The paper reviews seven expert systems shells in detail and indicates the strengths, weaknesses, and capabilities of each. The shells reviewed are IBM/AT compatible and range in price from $100 to $5000.

Palmer, R. N., and K. J. Holmes, Operational Guidance During Droughts: An Expert System Approach, *ASCE Journal of Water Resources Planning and Management Division*, ASCE, Vol. 114, No. 6, pages 647-666, 1988.

This paper describes a decision support system used to aid in drought management decisions for the Seattle Water Department. Its components include an expert system, a linear programming model, database management tools, and computer graphics. The knowledge base was developed through a series of interviews with Water Department personnel and augmented by over 2500 runs of the linear programming model. This system has been extensively used and verified by the Seattle Water Department.

Pazner, M. I., *Geographic Knowledge Base Design and Implementation*, Ph.D. Thesis, University of California, Santa Barbara, 1986 [University Microfilms order no. ADG86-14643].

Radford, A. D., Hung, P. and Gero, J. S., New Rules of Thumb from Computer-Aided Structural Design: Acquiring Knowledge for Expert Systems, In *Proceedings CADD-84*, Butterworth and Co., Ltd., U. K., 1984.

Pareto's optimization technique is proposed as an aid to the knowledge-acquisition process and illustrated using the floor system design as a paradigm.

Radford, A. D. and Gero, J. S., Toward Generative Expert Systems for Architectural Detailing, *Computer Aided Design*, Vol 17, No. 9, pages 428-434, November 1985.

Rank, E. and Babuska, I., An Expert System for the Optimal Mesh Design in the hp-Version of the Finite Element Method, *Numerical Methods in Engineering*, Vol. 24, pages 2087-2106, 1987.

Describes an intelligent assistant for modeling a hp version finite element program.

Rasdorf, W. and Wang, T., Generic Design Standards Processing in an Expert System Environment, *ASCE Journal of Computing in Civil Engineering*, Vol. 2, No. 1, pages 68-87, Jan. 1988.

A generic knowledge-based standard processing system has been developed by Rasdorf's research group. This paper presents an overview and implementation status of this system.

Reddy, D. V., Arockiasamy, M., Sriram, D., and Connor, J., Knowledge-Based Systems for Design and Analysis of Offshore Structures, In *ASME Conference on Computers in Artic Regions*, Feb. 1988.

Rehak, D., Expert Systems in Water Resource Management In James, W. and Torno, H. (editors), *Proceedings ASCE Conference on Emerging Techniques in Storm Water Flood Management*, Niagara on the Lake, Ontario, Canada, October 31 - November 4, 1983.

Current systems in water resource management are described.

Rehak, D., Howard, C., and Sriram, D., Architecture of an Integrated Knowledge Based Environment for Structural-Engineering Applications, In Gero, J. (editor), *Knowledge Engineering in Computer Aided Design*, pages 89-124, North-Holland, Elsevier Science Publishing Company, 1985.

A framework for distributed computer-aided engineering environment is provided. This framework is based on a distributed network of cooperating knowledge-based processing components.

Rehak, D., Christiano, P. P. and Norkin, D. D., Sitechar: An Expert System Component of a Geotechnical Site Characterization Workbench In Dym, C. L. (editor), *Applications of Knowledge-Based Systems to Engineering Analysis and Design*, American Society of Mechanical Engineers, pages 117-133, 1985.

A blackboard based system to determine the local geology of a site is described.

Rehak, D. and Lopez, L. A., *Computer-Aided Engineering: Problems and Prospects*, Civil Engineering System Laboratory Research Series 8,

University of Illinois, July 1981.

> Potential use of KBES for the development of an integrated structural design system is addressed.

Rehak, D. R. and Fenves, S. J., Expert Systems in Construction, In *Computers in Engineering 1984, Advanced Automation: 1984 and Beyond*, American Society of Mechanical Engineering, New York, pages 228-235, 1984.

> Discusses potential applications of expert systems to the construction industry.

Richardson, R. C.,*Intelligent Computer Aided Instruction in Statics*, Ph.D. Thesis, Colorado State University, 1986 [University Microfilms order no. ADG87-05480, Computer Science].

Richtie, S. G., Yeh, C., Mahoney, J., and Jackson, N. C., Surface Condition Expert System for Pavement Rehabilitation Planning, *ASCE Journal of Transportation Engineering*, Vol. 113, No. 2, pages 155-167, March 1987.

> Describes SCEPTRE, which assists highway engineers in planning cost-effective flexible pavement rehabilitation strategies. SPECTRE runs on an IBM-PC and is implemented in EXSYS^{tm}.

Rivlin, J. M., Hsu, M. B. and Marcal, P. V., *Knowledge-based Consultant for Finite Element Analysis*, Technical Report AFWAL-TR-80-3069, Flight Dynamics Laboratory (FIBRA), Wright-Patterson Air Force Base, May 1980.

> A KBES implemented in FORTRAN and interfaced to the MARC non-linear analysis program.

Roddis, W.M.K., and Connor, J., *Qualitative/Quantitative Reasoning for Fatigue and Fracture in Bridges*, Coupling Symbolic and Numerical Computing in Expert Systems II, edited by Kowalik, K. and Kitzmiller, C.T., North-Holland, 1988.

> A multi-level framework for fatigue and fracture assessment of bridges is presented.

Rosenman, M. A. and Gero, J. S., Design Codes as Expert Systems, *Computer-Aided Design*, Vol. 17, No. 9, pages 399-409, Nov. 1985.

> Shows how expert systems can be used to check designs against various standard codes.

Rosenman, M. A. et al., Expert Systems Applications in Computer-Aided Design, In Smith, A. (editor), *Knowledge Engineering and Computer Modeling in CAD*, pages 218-225, Butterworths, London, 1986 [see also *Computer Aided Design*, Vol. 18, No. 10, pages 546-551, December 1986].

> Expert systems to do retaining wall design and the design of kitchens are discussed.

Rosenman, M. R., Balachandran, B. M. and Gero, J. S., The Place of Expert Systems in Civil Engineering, In R. Hadgraft and W. Young (editors), *Symposium on Knowledge-Based Systems in civil Engineering*, Monash University, Clayton, pages 19-36, 1988.

> This paper examines the potential place of expert systems in civil engineering and presents a number of expert systems.

Soh, C, and Soh, A., Example of Intelligent Structural Design System, *ASCE Journal of Computing in Civil Engineering*, Vol. 2, No. 4, pages 329-345, Oct. 1988.

> ISTRUDS, a framework for knowledge-based structural design, and IPDOJS, an example of a preliminary design of fixed-steel offshore jacket structures, are described.

Skibniewski, M., Framework for Decision-Making on Implementing Robotics in Construction, *ASCE Journal of Computing in Civil Engineering*, Vol. 2, No. 2, pages 188-201, April 1988.

> The role of KBES for construction robotics is explored and a knowledge-based framework is presented.

Slater, J. H., Petrossian, R., and Shyam-Sunder, S., An Expert Tutor for Rigid Body Mechanics - MACAVITY, In *Proceedings of the Expert Systems in Government Symposium*, Edited by K. Karna, IEEE Computer Society, pages 24-35, 1985.

> Describes work at M.I.T. on intelligent computer aided instruction.

Smith, R. G. and Baker, J. D., The Dipmeter Advisor System: A Case Study in Commercial Expert System Development, In *Proceedings Eighth IJCAI*, Morgan Kaufmann Publishers, pages 122-129, 1983 [See also Davis, R. et al., The Dipmeter Advisor: Interpretation of Geologic Signals, In Proceedings Seventh IJCAI, pages 846-849, 1981].

> DIPMETER ADVISOR is an interactive interpretative system used for inferring subsurface geologic structure using *dipmeter*. The *dipmeter* is a tool to measure the petrophysical properties of the subsurface. Specifically, it measures the conductivity of the rock in a number of directions and uses this information with the inclination and orientation of the tool to estimate the magnitude and azimuth of the tilt of various formation layers penetrated by a borehole.

The knowledge-base consists of rules that infer the existence of geologic structures from the presence of various features in a log. Additionally, there are feature detection procedures used to scan the log data which provide the initial level of data reduction. The control mechanism is a data-driven inference mechanism which matches patterns in the database with appropriate rules. DIPMETER advisor is implemented in STROBE, a frame-based language.

Sriram, D., Maher, M., Bielak, Fenves, S. J., *Expert Systems for Civil Engineering - A Survey*, Technical Report R-82-137, Department of Civil Engineering, Carnegie-Mellon University, June 1982.

> Written as an introduction to KBES for civil engineers. A number of current expert systems, KBES building tools and potential applications in structural and geotechnical engineering are described.

Sriram, D., Maher, M. and Fenves, S., Applications of Expert Systems in Structural Engineering, In *Proceedings Conference on Artificial Intelligence*, Oakland University, Rochester, MI, pages 379-394, April 1983 [For an extended version see *Computers and Structures*, January 1985].

> Discusses the applications of KBES to various phases of structural design.

Stanford, G., *Potential Applications of Expert Systems in Geotechnical Engineering*, Master's Thesis, Department of Civil Engineering, Carnegie-Mellon University, April, 1983.

> Potential applications in geotechnical engineering, specifically in the domain of Landslide engineering, are addressed. The author also relates his experience of knowledge acquisition from a domain expert and from literature.

Thompson, J. et al., Extending Prolog to Provide Better Support for Design Code Expert Systems, *Microcomputers in Civil Engineering*, pages 93-109, Vol. 3, No. 2, 1988.

> The CSIRO Division of Building Research KBS Group in Australia has developed a rule-based shell - Bx-Shell - in PROLOG. Extensions that increase the run time efficiency of Bx-Shell and a KBES called WINDLOADER, which assists the engineer in determining the wind loads on a structure, are described.

Thomson, J., et al., An Expert System to give Advice to Operators of a Metropolitan Water Supply, Drainage and Sewerage Network, In Gero, J. and Stanton, R. (editors),*Artificial Intelligence Developments and Applications*, Elsevier Science Publishers, pages 49-64, 1988.

> Describes a collaborative development project between CSRIO and the Melbourne and Metropolitan Board of works. The problem of building a KBES for monitoring the water and sewage supply has many interesting characteristics, e.g., there are many installations of treatment plants that are networked. The system is implemented in a PROLOG-based KBES shell (see discussion in the previous reference).

Topping, B. H. V. and Kumar, B., The Application of Artificial Intelligence to Civil and Structural Engineering: A Bibliography, In *Artificial Intelligence: Techniques and Applications for Civil and Structural Engineering*, Civil-Comp Press, Edinburgh, 1989.

> A bibliography on AI applications in civil engineering. The bibliography is divided into the following sections: Books: General; Books: Civil and Structural Engineering; Bibliographies in Engineering; General Reports; General Papers: Civil and Structural Engineering; Expert System Shells; Logic Programming, Interfaces and Knowledge Elicitation; Uncertainty; Databases; Constraints, Codes, Rules, and Design; Surveying and Road Layout; Architecture, Layout, Space, and Building Layout; Civil Engineering; Construction Planning, Management and Control; Structural Engineering; Structural Analysis and Design; Offshore Engineering; Structural Optimization; Finite Element Analysis and Modelling; Water Engineering; Environmental Engineering; Geotechnical Engineering; Education; Materials Engineering; Pavement Engineering; and Transport Engineering.

Tung, S. I., *Designing Optimal Transportation Networks: A Knowledge-Based Computer-Aided Multicriteria Approach*, Ph. D. Thesis, University of Washington, 1986. (University of Microfilms Order Number ADG86-26709)

Turkiyyah, G. and Fenves, S., Getting Finite Element Programs to Reason about their Assumptions, In Nelson Jr., J. K. (editor), Computer Utilization in Structural Engineering, ASCE, pages 51-60, 1989.

> Most of the papers on intelligent front ends to analysis programs deal with shallow (heuristics) knowledge. This paper takes a view that causal (deep) reasoning should alos be incorporated into these intelligent front ends.

Wager, D. M., Can Expert Systems Help the Construction Industry?, In *CIVIL-COMP 85. Proceedings of the Second International Conference on Civil and Structural Engineering Computing*, Civil-Comp Press, 10 Saxe-Coburg Place, Edinburgh, U.K., pages 387-389, Dec. 3-5, 1985.

Weiss, S. M. and Kulikowski, C. A., Building Expert Programs for Controlling Complex Programs, In *Proceedings 2nd NCAI*, pages 322-326, 1982.

A KBES for well log analysis is described.

Wijesundera, D. A. and Harris, F. C., The Integration of an Expert System into the Construction Planning Process, In *CIVIL-COMP 85, Proceedings of the Second International Conference on Civil and Structural Engineering Computing*, London, England, pages 399-405, Dec. 3-5, 1985.

Zhao, F. and Maher, M. L., Using Analogical Reasoning to Design Buildings, *Engineering with Computers*, Vol. 3, No. 4, pages 107-114, 1988.

STUPLE uses transformational analogy to design new buildings, based on previous designs.

Zhang, Z. and Simaan, M., A Rule-Based Interpretation System for Signal Images, In *Proc. SPIE Int. Soc. Opt. Eng.*, SPIE, P. O. Box 10, Bellingham, Washington 98227, Vol. 635, pages 280-287, 1986.

Texture based techniques are used in the interpretation of seismic images.

Zimmie, T., Law, K., and Chapman, D., Development of a Rule Based Expert System for Characterization of Harardous Waste Sites, In *Proceedings of the ASCE National Conference on Environmental Engineering*, July 1986.

Describes a KBES, implemented in OPS5, that facilitates the investigation of inactive hazardous waste sites.

Zozaya-Gorostiza, C. and Hendrickson, C., An Expert System for Traffic Signal Setting Assistance, *Journal of Transportation Engineering*, ASCE, Vol. 113, No. 2, pages 108-126, 1987.

Describes an experimental knowledge-based expert system to assist in traffic signal setting for isolated intersections. The system, TRALI, can be applied to highly irregular geometries. The programming environ- ment is OPS5.

3 BOOKS AND EDITED COLLECTIONS

Adeli, H. and Balasubramanyam, K., *Expert Systems for Structural Design: A New Generation*, Prentice-hall, Inc., 1988.

Describes the work on BTEXPERT, which is a KBES for optimum design of bridge trusses. Current version of BTEXPERT, implemented as a coupled system (symbolic and numerical), can design four types of bridge trusses: Pratt, Parker, Parallel chord K truss, and Curved-chord K truss for a range of 100 to 500 ft.

Adeli, H., *Microcomputer Knowledge-Based Expert Systems in Civil Engineering*, ASCE, 1988.

Contains papers presented in a symposium. Papers are organized into structural engineering (8), geotechnical and environment engineering (3), construction (3), and general (2). Structural engineering papers are: Knowledge Acquisition and Machine Learning in an Expert System (Zhang and Yao), Expert System RAISE-1 (Chen and Liu), Using PROLOG on a Macintosh to Build an Engineering Expert System (Roddis and Connor), SDL: An Environment for Building Integrated Structural Design Expert Systems (Paek and Adeli), An Integrated Rule-Based System for Industrial Building Design (Kumar and Topping), A Rule-Based System for Estimating Snow Loads on Roofs (Fazio, Bedard and Gowri), PC PLUS + LOTUS 123 (Malasri), A Knowledge-based Expert System for the Selection of Structural Systems for Tall Buildings (Jayachandran and Tsapatsaris). Geotechnical papers are: GEOTOX-PC: A New Hazardous Waste Management Tool (Mikroudis and Fang), An Expert System for Diagnosis and Treatment of Dam Seepage Problems (Asgian et al.), ASA: An Expert System for Activated Sludge Analysis (Parker and Parker). Construction papers are: Knowledge Elicitation Techniques for Construction Scheduling (De La Garza et al), An Expert System for Construction Claims (Kim and Adams), Knowledge Engineering in a Knowledge-Based System for Contractor Prequalification (Russell and Skibniewski). General papers are: Logic Programming to Manage Constraint based Design (Chan and Paulson, Jr), Expert Systems for the Earthquake-Related Industry (Kim et al.).

Adeli, H. (editor), *Expert Systems in Construction and Structural Engineering*, Chapman and Hall, London, UK, 1988.

Contains 15 articles on applications of KBES in civil engineering. First five chapters are contributed by Adeli. The other chapters are: Expert Systems for Structural Design (Maher, Fenves and Garrett), Expert Systems Applications in Construction Engineering (Finn), Knowledge Engineering for a Construction Scheduling Analysis System (Ibbs and De La Garza), Approximate Reasoning in Structural Damage Assessment (Ross), Condensation of the Knowledge Base in Expert Systems with Applications to Seismic Risk Evaluation (Dong et al.), Expert Systems for Condition Evaluation of Existing Structures (Zhang and Yao), An Expert System for

Earthquake Intensity Evaluation (Liu and Wang), A Knowledge-based Approach to Engineering Information Retrieval (Rasdorf and Watson), Knowledge Acquisition for Expert Systems in Construction (Trimble), and Codes and Rules and their Roles as Constraints in Expert Systems for Structural Design (Fukuda).

Akman, V., teh Hagen, P. J. W., and Veerkamp, P. J. (editors), *Intelligent CAD Systems II: Implementation Issues*, Springer-Verlag New York, Inc., 175 Fifth Avenue, New York, NY 10160-0266, 1989 [See also Volume I].

> Contains a number of papers presented at the Second Eurographics Workshop on Intelligent CAD. Papers in this volume emphasize on implementation issues.

Coyne, R., *Logic Models of Design*, Pitman Publishing, 128 Long Acre, London WC2E 9AN, 1988.

> Discusses the role of logic in modelling design tasks. The distinction between knowledge and reasoning is clarified. The various perspectives of design are discussed. Inference strategies for planning and design problems are described. The book also addresses the algorithmic complexity of design problems. This is a well written book and is a recommended reading for people working in design automation.

Dym, C. L. (editor), *Applications of Knowledge-Based Systems to Engineering Analysis and Design*, American Society of Mechanical Engineering, 1985.

> Contains several papers on the applications of KBES in civil engineering and mechanical engineering.

Gero, J. (editor), *Knowledge Engineering in Computer-Aided Design*, North-Holland Publishing Company, 1985.

> Contains a number of papers on the application of KBES in engineering. Papers encompass civil, architecture, mechanical and electrical engineering applications. Papers relevant to civil engineering are: Requirements and Principles for Intelligent CAD Systems (Tomiyama and Yoshikawa), Conceptual Design of CAD Systems Involving Knowledge Bases (Ohsuga), Knowledge-Based Computer-Aided Architectural Design (Gero, et al.), Architecture of an Integrated Knowledge Based Environment for Structural Engineering Applications (Rehak, et al.), HI-RISE: An Expert System for the Preliminary Structural Design (Maher and Fenves), Knowing where to Draw the Line (Szalapaj and Bijl), CAD, KE and Ada (Mac an Airchinnigh), Encoding Requirements to Increase Modelisation Assistance (Cholvy and Foisseau), and An Object-Centered Framework for Expert Systems in Computer-Aided Design (Barbuceanu).

Gero, J. (editor) *Expert Systems in Computer Aided Design*, North-Holland Publishing Company, 1987.

> An excellent collection of papers on applications of KBES for design. The discussions accompanying each paper makes the book very interesting.

Gero, J. (editor), *Artificial Intelligence in Engineering: Robotics and Processes*, Computational Mechanics Publications, UK, 1988.

> Papers in this book are grouped into: robotics, geometric and spatial reasoning, interpretation processes, reasoning processes, planning and scheduling, and interfaces.

Gero, J. (editor), *Artificial Intelligence in Engineering: Design,* Computational Mechanics Publications, UK, 1988.

> Contains papers related to engineering design, which were presented at the Third International Conference on the Applications of AI in Engineering Problems. Papers are grouped into: design knowledge and representation, integrated circuit design, mechanical engineering design, structural engineering design, simultaneous engineering design, and architectural design.

Gero, J. (editor), *Artificial Intelligence in Engineering: Diagnosis and Learning,* Computational Mechanics Publications, UK, 1988.

> Contains papers presented at the Third International Conference on the Applications of AI in Engineering Problems. These papers are grouped into: diagnosis from structure and behavior, integrated diagnostic reasoning, diagnosis as control, diagnosis processes and environments, and learning and tutoring.

Giraud, C., (editor), *CAD and Robotics in Architecture and Construction*, Proceedings of the Joint International Conference at Marseilles, 25-27 June 1986, Kogan Page, London (Nichols Publishing Company, New York), 1986.

> Papers include: AI in Architectural CAD (Bijl), A Schematic Representation of the Designers' Logic (Akin et al.), Knowledge-based Computer Aided Design (Kalay et al.), Modelling Design Descriptions (Krishnamurthi), ROOSI- Version One of a Generative Expert System for the Design of Building Layouts (Flemming et al.), Expert Systems in Construction: Initial Experiences (Kosela, et al.), Geometry and Domain Modelling for Construction Robots (Woodbury, et al.), etc.

Kostem, C. and Maher, M. (editors), *Expert Sys-

tems in Civil Engineering, ASCE, 1986.

> Contains several papers from a ASCE workshop on expert systems in civil engineering. Structures related papers are: DAPS: An Expert System for Damage Assessment of Protective Structures (Ross, et al.), Design of an Expert System for Rating of Multibeam Highway Bridges (Kostem), Seismic Risk Analysis Systems (Miyasato, et al.), Design of a Knowledge based System to Convert Airframe Geometric Models to Structural Models (Gregory and Shephard), RC Structures under Severe Loads - An Expert Systems Approach (Krauthammer and Kohler), Knowledge Engineering in Object and Space Modeling (Aikner), Qualitative Physics and the Prediction of Structural Behavior (Slater), Use of Knowledge-Based Expert Systems in Seismic Analysis of Structures (Lashkari and Boissonnade), Developments in Expert Systems for Design Synthesis (Gero and Coyne). Construction related papers are: Expert Systems in an Engineering Construction Firm (Finn and Reinschmidt), Howsafe: A Microcomputer-Based Expert System to Evaluate the Safety of a Construction Firm (Levitt), An Expert System for Construction Schedule Analysis (O'Connor, et al.), Application of Expert Systems to Construction Management Decision-Making and Risk Analysis (Kangari), An Expert System for Site Selection (Findikaki), Applications of Automated Interpretation of Data (Maser). Environmental related papers are: Fault Diagnosis of Hazardous Waste Incineration Facilities using a Fuzzy Expert System (Huang, et al.), An Expert System for Flood Estimation (Fayegh and Russell), and Selection of Appropriate Water/Sanitation Treatment Technology in Developing Countries: An Expert System Application (Arnold). General papers are: What is an Expert System (Fenves), Problem Solving using Expert System Techniques (Maher), Expert System Tools for Civil Engineering Applications (Ludvigsen, et al.), Attributes and Characteristics of Expert Systems (Kostem), Expert Systems in an Engineering Firm (Finn and Reinschmidt), and KBES and Interactive Graphics (Vora).

Latombe, J-C. (editor), *Artificial Intelligence and Pattern Recognition in Computer Aided Design*, North-Holland Publishing Company, New York, 1978.

> Contains a number of pioneering papers on the application of AI to engineering design problems.

Maher, M., (editor) *Expert Systems for Civil Engineers: Technology and Application*, ASCE, 345 East 47th St., NY 10017, 1987.

> This book is divided into two parts. First part contains three chapters: Expert systems components, Languages and tools for building expert systems, and Implementation issues in the building of expert systems. Part II consists of applications in Structural, Geotechnical, Construction, Environmental, and Transportation engineerings. Over 70 systems are reported.

Nelson, Jr., J. K. (editor), *Computer Utilization In Structural Engineering*, American Society of Civil Engineers, 1989.

> Consists of papers presented at Structures Congress '89. Papers relevant to knowledge-based systems are: Getting Finite Element Programs to Reason about their Analysis Assumptions (Turkiyyah and Fenves), Representation of Strategic Choices in Structural Modeling (Dym and Salata), Idealized Models n Engineering Analysis (Shephard, Horngold, and Wentorf), Object-Oriented Design of Finite Element Programs (Baugh and Rehak), An Object-Oriented Model for Building Design and Construction (Garrett, et al.), A KBES for Seismic Design of Buildings (Subramani, et al.), A Prototype Expert System for Steel Connections (Gu, et al.), Automated Design of Reinforced Concrete Components (Quimby and Balling), BABE: An Expert System for Structural Design of Bridge Abutments and Piers (Zheng, et al.), An Integrated System for Seismic Vulnerability and Risk for Engineering Facilities (Dong, et al.), QSEIS: A Knowledge Based System for Deriving Qualitative Seismic Behavior (Ganguly, et al.), and A Knowledge Based System for Evaluating the Seismic Resistance of Existing Buildings (Fenves and Ibarra-Araya).

Niwa, K., *Knowledge-Based Risk Management in Engineering*, Wiley Series in Engineering and Technology Management, Wiley Interscience, John Wiley & Sons, 1989.

> Emphasis is on the risk management of large construction projects. Presents the standard work package method, which can be viewed as a mapping between project risks and project work elements.

Palmer, R. N., Special Issue of the ASCE Journal of Computing in Civil Engineering, Vol. 1, No. 4, 1987.

> Papers include: Software for Expert Systems Development (Ortolano and Perman), Expert Systems for Geotechnical Engineers (Santamarina and Chameau), Expert System for Construction Planning (Hendrickson, et al.), Expert Systems for Structural Design (Maher), Expert System for Drought Management Planning (Palmer and Tull), New Expert Systems in Environmental Engineering (Ortolano and Steinemann), Expert Systems in Construction: Work in Progress (Ashley and Levitt), Expert Systems in Structural Engineering: Works in Progress (Allen).

Pham, D. T. (editor), *Expert Systems in Engineering*, IFS Publications/Springer-Verlag, 1988.

Contains 29 papers on the applications of KBES in engineering. The book is organized into six chapters: Introduction to Expert Systems, Expert Systems in Process Engineering, Expert Systems in Civil Engineering, Expert Systems in Electrical and Electronic Engineering, Expert Systems in Mechanical Engineering, Expert Systems in Manufacturing Engineering. Most of the papers are reprints from various sources. Overall the collection is good.

Rychener, M. D. (editor), *Expert Systems for Engineering Design*, Academic Press, 1988.

This book is a collection of papers on engineering design work done at Carnegie Mellon University's Engineering Design Research Center. The first chapter is an introduction to expert systems in engineering design. The rest of the book is divided into three parts: Synthesis, The Nature of Expertise, and Integrated Software Organizations. The part on Synthesis includes five papers: HI-RISE: An Expert System for Preliminary Structural Design, The DECADE Catalyst Selection System, Rule-Based Systems in Computer-Aided Architectural Design, Single Board Computer Systems, and Knowledge-Based Alloy Design. The part titled "The Nature of Expertise" consists of two papers: Expertise of the Architect, and A Graphical Design Environment for Quantitative Decision Models. Integration part is comprised of four papers: Artificial Intelligence Techniques: Expanding VLSI Design Automation Technology, Building Large Scale Software Organizations, ARCHPLAN: An Architecture Planning Front End to Engineering Design Expert Systems, and Design Systems Integration in CASE.

Sriram, D. and Adey, R. (editors), *Applications of Artificial Intelligence in Engineering*, Springer Verlag/CM Publications, U. K., 1986.

Contains about 100 papers on the applications of artificial intelligence to engineering problems.

Sriram, D. and Adey, R. (editors), *Artificial Intelligence in Engineering: Tools and Techniques*, Computational Mechanics Publications, U. K., 1987.

Sriram, D. and Adey, R. (editors), *Knowledge-Based Expert Systems for Engineering: Classification, Education, and Control*, Computational Mechanics Publications, U. K., 1987.

Sriram, D. and Adey, R. (editors), *Knowledge-Based Expert Systems in Engineering: Planning and Design*, Computational Mechanics Publications, U. K., 1987.

The above three books are based on the Second International Conference on AI in Engineering. They contain around 80 papers.

Sriram, D., *Knowledge-Based Approaches for Structural Design*, Computational Mechanics Publications, Southampton, 1987.

This book describes DESTINY, developed for integrated structural design, and ALL-RISE system, developed for preliminary synthesis of buildings. These systems are compared with other leading engineering design systems, which incorporate artificial intelligence concepts.

Topping, B. H. V. (editor), *The Application of Artificial Intelligence Techniques to Civil and Structural Engineering*, Civil-Comp Press, 10 Saxe-Coburg Place, Edinburgh, U.K., 1987 [See also Civil-Comp 1989].

Contains papers presented at CIVIL-COMP 85 and CIVIL-COMP 87. The book is divided into the following sections: 1) Expert System Shells, Interfaces and Knowledge Elicitation, 2) Construction Planning, Management and Control, 3) Structural Analysis and Design, 4) Water Engineering, 5) Geotechnical Design, 6) Materials Engineering, 7) Pavement Design, and 8) Education.

Will, K., (editor) *Computing in Civil Engineering: Microcomputers to Supercomputers*, Proceedings of the Fifth Conference on Computing in Civil Engineering, ASCE, 1988 [Several papers on KBES also appear in the Proceedings of the Sixth Conference on Computing in Civil Engineering, ASCE, September 1989].

Contains several papers related to KBES. Structures related papers are: Grammers for functional and Spatial Reasoning in Design (Fenves and Baker), Knowledge Representation and Processing for Computer Integrated Structural Design (Sause and Powell), KADBASE: An Expert System/Database Interface (Howard), An Integrated Software Environment for Building Design and Construction (Fenves et al.), An Expert System for the Preliminary Design of Frameworks (Ovunc), A Knowledge-based Preprocessor for Nonlinear baseplate Analysis (Al-Shawaf and Pauschke), Object Representations for Structural Analysis and Design (Fenves, G.). Construction related systems are: An Expert System for Selecting Bid Markups (Ahmad and minkarah), An Expert System for Contractor Prequalification (Russell and Skibniewski), CONSITE: A Knowledge Expert System for Site Layout (Hamiani and Popescu), Expert System for Management of Low Volume Flexible Payments (Aougab, et al.). Water resources related systems are: An Expert Advisor for the Qual2e Water Quality Model (Barnwell et al.), A Parameter Estimation Expert System for the USGS Modular Groundwater Model (Lennon et al.), An Expert System Approach to Hydrologic Data Fusion (Scarlatos).

Zozaya-Gorostiza, C. and Hendrickson, C., and

Rehak, D., *Knowledge-Based Process Planning for Construction and Manufacturing*, Academic Press, April 1989.

> Presents PLANEX, a domain independent knowledge-based process planning framework. PLANEX has been used in developing KBES for construction and manufacturing problems.

Chapter 14
Hardware and Software for Expert Systems Courses

by Phillip Ludvigsen, A.M. ASCE and Mary Lou Maher, A.M. ASCE

1 INTRODUCTION

This chapter addresses important considerations related to selecting appropriate hardware and software to be used by students when building a small expert system as an exercise. The chapter generalizes rather than cites specific tools. The review of all hardware and software tools is an immense and always incomplete task. Rather than attempt such a review, the purpose of this chapter is to provide guidance in the form of important considerations and references to publications that review specific tools. This chapter begins with a discussion of the characteristics of expert system coursework, followed by considerations for hardware and software.

2 EXPERT SYSTEM COURSEWORK

Building expert systems as a pedagogical exercise in an expert system course differs tremendously from large-scale development of industrial systems. Important industrial considerations such as processing speed and the ability to integrate large external data bases are secondary to factors that allow the student to build and disseminate various small expert systems quickly and at little or no cost.

A course on building engineering expert systems can not afford large allocations of class time to the logistics of hardware and software utilization. Students don't have sufficient time to learn the intricacies of a specialized programming environment running on dedicated hardware (e.g., Lisp Machines). The instructor should select commonly utilized computers as well as software that can be applied easily to fundamental engineering problems. Time constraints on the instructor also dictate selection of problem domains that are quickly understood by the students. Trying to teach the principals of building engineering expert systems in addition to teaching complex engineering concepts can be a tremendous task.

Most universities cannot financially afford to accommodate a large number of students with dedicated LISP Machines running sophisticated Artificial Intelligence (AI) development software. A more realistic scenario involves utilizing existing shared computer resources with multipurpose, and usually less expensive, software such as spreadsheets (e.g., Lotus 1-2-3^{TM}), conventional computer languages (e.g., Pascal or C), specialized AI languages (e.g., Prolog or LISP), and purchasing special purpose expert system shells (e.g. LEVEL5^{TM} or VPExpert^{TM}). If multiple copies of an expert system shell are within the budget, the liability for "run-time" licensing should be investigated. It is important that students be able to legally take the results of their course work with them upon completion of the course.

3 HARDWARE

Given the typical constraints of available participant time and computer resources that affect many engineering laboratory courses, it is reasonable to expect available hardware to be used for multiple purposes. In this light, a major consideration is that the hardware be accessible for use by other groups outside of those interested in just building expert systems. In this section we consider hardware in three categories: mainframes, mini and microcomputers, and Lisp machines.

Mainframe computers have traditionally offered centralized control of data security, full-time technical support, and maintenance. Students usually have easy access (day or night) from remote locations. Mainframe use can also be cost effective since only one copy of software can service an entire class. Although mainframes can be convenient, access tends to be limited to character-based terminals, which means some specialized programming environments (e.g., KEE^{TM}) can not be used. Even AI languages (e.g., Lisp) that do not require a graphics terminal can use a lot of core memory thus increasing the time it takes to process a given task. This explains why some schools only allow AI languages and programming tools to be run on their Mainframes at night.

Single or multi-user mini and microcomputers

offer many of the same benefits as large mainframes. These machines have the potential for centralized control and remote access. On the other hand, providing comprehensive support and maintenance could be overwhelming for the instructor. Day to day upkeep of the system may be distributed among the students, but this additional burden may cause distraction from the main objective of learning about expert systems.

Shared computer resources usually take the form of multi-user systems or numerous personal computers networked together. There are many variations between these two forms. In the past, multi-user systems implied large mainframes accessible via terminals connected directly or by phone modem. Today multi-user capabilities as well as parallel processing can be found on personal computers. The distinction between mainframes and smaller distributed systems is quickly becoming unnoticeable by users.

One last area of hardware that may be appropriate for advanced classes is the LISP machines. LISP machines fall into a very specialized hardware category that has been optimized to run the LISP programming language. It is unlikely, however, that novice students of a one term expert system course would require the speed and flexibility of a sophisticated LISP machine. It is equally unlikely that an engineering program with a tight budget could afford $30,000 to $50,000 per single user machine. Within the last few years, the decline in LISP machine sales raises doubts to the utility of this type of hardware for even large-scale expert system efforts.

The most inexpensive hardware option is the microcomputer. Many students either own their own or have access to microcomputers on campus for other courses. The memory and speed limitations that may indicate the need for a more powerful machine are not critical issues for a class project. Portability, i.e. students can carry their program to another computer on a floppy disk, is an advantage.

4 SOFTWARE

Various reviews of software for building expert systems, mainly for microcomputers, have been published in the last three years (Harmon 1986, Ludvigsen et al. 1986, Richer 1986, Freeman 1987, Ludvigsen and Grenney 1987, Olsen et al. 1987, Ortolano and Perman 1987, Somsel 1987, Mullarkey 1987, Buynjolfsson and Loofbourrow 1988, Rasmus 1988). These reviews present perceived strengths and weaknesses as well as various perspectives into selecting an appropriate tool. Many of these reviews carry a useful listing of currently available tools and prices. It is important to note that few areas of software development have shown as high a rate of growth as expert system tools. Some tools have undergone two or three generations of development since they were first introduced, thus one should be aware of current changes in program features and price.

Brynjolfsson and Loofbourrow (1988) group current expert system tools into four broad categories: inductive tools, simple rule tools, hybrid tools, and languages. Inductive tools build an IF/THEN structure from examples of accepted solutions. Simple rule tools apply IF/THEN rules entered by the knowledge engineer. More complex features such as a graphical interface and various knowledge representations are typical of hybrid tools. Languages such as Pascal, C, Prolog, and LISP allow knowledge engineers to develop customized knowledge representation, inferencing, and interfaces for complex problem domains.

Mullarkey (1987) groups software for developing expert systems into three categories: general purpose programming languages, general purpose representation languages, and expert system frameworks or shells. The general purpose programming languages include Lisp, Pascal, and C. The use of one of these languages requires the student implement his own inferencing scheme and knowledge base representation structures. The general purpose representation languages are derived from AI research and include KEE™, Knowledge Craft™, and Prolog. The use of a representation language provides the student with a structure for developing the knowledge base but often requires that the student implement his own user interface and inferencing scheme. The expert system shells include EXSYS™ and Deciding Factor™. These are the easiest of the three categories of software to learn to use. They incorporate many of the concepts of expert systems in the package so the student does not need to write an inference mechanism to support a knowledge base structure.

5 SELECTING AN APPROPRIATE TOOL

Selecting an appropriate tool is an important early step in the development of any expert system. It is a decision that depends on who is going to use the tool and how much knowledge engineering experience they possess. Citrenbaum et al. (1987) propose four user categories (Student, Domain Expert, Knowledge Engineer, and Developer) and five possible activities (Learning, Knowledge Engineering, Initial Prototype, Expanded Prototype, and Delivery System) involved in expert systems development. For course work, the user would be placed into the Student category during a Learning stage of expert systems development. For the Student user in the Learning stage, Citrenbaum and his co-authors list the most important features as intuitive knowledge representation(s); low-cost hardware/software platform; casual user interface; and user default values. These considerations are similar to those in selected appropriate hardware in that they (1) expedite the learning process and (2) reduce costs.

No single software package is best for all expert system courses. Ease of use is of paramount interest so that students may concentrate on the concepts rather than on the syntax of a particular programming language. Of almost equal interest is a match between the representation paradigm(s) incorporated in the software package and those being taught. The course content may include rule based representation, frame based representation, reasoning with inexact knowledge, logic programming, induction, etc. Finding an easy to use software package that covers all these areas is impossible, and it is unreasonable to expect a student to learn more than one tool in a single term. The most popular option is an inexpensive rule based package that can accommodate uncertainty and can access other programs.

The issues of the learning curve for expert system tools and cost of the software are considered further.

5.1 Learning Curve

An important feature that affects how fast a student moves up the learning curve is the tool-user interface. The first tools had very little or no developer interface. Today, most tools come with at least an editor, trace and debug facilities as well as some kind of report and screen formatting capability. In general, the more graphical the interface, the easier the tool is to work with. Graphical interfaces are standard features for most upper end (above $5000) tools and LISP machines languages.

Many low end tools are starting to utilize pull-down menu and "point and click" technology. Expert system tools and languages that run on the Macintosh^{TM} family of microcomputers incorporate a graphical interface that is inherent in the system itself. A graphical interface can be extremely useful in visualizing the forest of knowledge amongst the individual trees of information.

Another mechanism for assisting novice knowledge engineers is automatic rule formation. It is a standard feature of many low end microcomputer-based tools. One popular technique for automatic rule formation is based on the "Iterative Dichotomizer 3" (ID3) algorithm developed by J. Ross Quinlan in 1979. To use this algorithm, the student would list key factors, corresponding descriptions, and conclusions of solved problems into a matrix. The computer then generates an optimized decision tree which is translated into one large series of nested IF-THEN-ELSE statements. Students can learn how to structure a small knowledge base by observing the results of the algorithm. Unfortunately on large problems, the resulting knowledge base can be difficult to follow and understand (Cohen and Feigenbaum, 1982).

5.2 Cost

Purchasing multiple copies of expert system development software can be an expensive proposition. Several options exist for reducing the cost to the institution.

1. Choose an inexpensive shell and have the students buy their own copy.

2. Buy an expensive package and have the students time share.

3. Look for academic discounts.

Some tools are tiered in sophistication and cost. For example, there may be a student version and a professional version of the tool. The student version may limit the size of the knowledge base or access to external programs, but meet the students' needs at a reduced cost. In most cases, the student version can be upgraded to a professional version for additional cost if the student is interested later. The more expensive professional versions can be reserved for smaller advanced expert system courses or thesis development.

Mainframe and networking versions of development tools may be cost effective for very large class sizes. This software, however, tends to cost much more than the equivalent single user versions (up to 10 times as much). Alternatively, many schools do not carry expert systems hybrid tools on their mainframes, but support languages such as LISP and Prolog.

Network file servers are notoriously capable of disseminating multiple copies of executable code directly into the core memory of other compatible computers. These computers can then execute the program as if copied from a diskette. Networking differs from time sharing: time sharing, one computer with one CPU, shares one copy of executable code with several terminals, where networking provides actual or virtual copies of executable code to each computer in the network. The practice of time sharing microcomputer software seemed to be influencing software developers to alter their standard licensing agreements. In the past, it was common to see an agreement that required buyers to purchase one copy of software per machine (or CPU). Today we see confusing contracts that awkwardly address fundamental, albeit technical, difference between computer time sharing versus networking.

A common method for Universities and Colleges to control software costs is to take advantage of academic discounts. Many tool manufacturers offer academic discounts in an effort to gain prestige from highly visible research projects while building a future user base. Academic discounts can be as much as 90 percent or more. Another method for lower software costs is to use public domain programs. Various expert system tools and languages for PCs and even mainframes are available from electronic bulletin boards and shareware catalogs. This software, however, tends to have poor or no documentation, program errors and more potential viral infections.

6 CONCLUSIONS

The selection of hardware and software for an expert system course depends on many factors, the most important ones are ease of use for the student and cost for the institution. This chapter discusses these issues with reference to currently available technology. The potential instructor of an expert system course is encouraged to read the referenced publications for information about specific tools. This paper indicates some of the most commonly used hardware and software platforms for expert system coursework, but the choice depends very much on the course content and the available resources.

7 REFERENCES

Brynjolfsson E. and Loofbourrow T., An Overview of Expert System Building Tools for PCs, *PC AI*, September/October, 1988, pp. 36-43.

Citrenbaum, R., Geissman, J.r., and Schultz R. Selecting a Shell, *AI-Expert*, September, 1987, pp. 30-39.

Cohen, P.R. and Feigenbaum E.A, Eds. *The Handbook of Artificial Intelligence, Volume 3*, HeurisTech Press, Stanford, CA., 1982, p. 639.

Freedman, R., 27-Product Wrap-up Evaluating Shells, *AI-Expert*, September, 1987, pp. 69-74.

Harmon, P., The Cost-Effectiveness of Tools, *AI-Expert*, September, 1986, pp. 69-74.

Ludvigsen, P.J., Grenney W.J., Dyreson D., and Ferrara J.M., Expert System Tools for Civil Engineering Applications, *Expert Systems in Civil Engineering Symposium Proceedings*, ASCE, New York, NY., 1986, pp. 18-29.

Ludvigsen, P.J. and Grenney W.J., The Shell Game: Expert System Tools for Civil Engineering Application - Part II, *Computing in Civil Engineering Proceedings of Fifth Conference*, ASCE, Alexandria, VA. March, 1987, pp. 361-170.

Mullarkey, P.W., Languages and Tools for Building Expert Systems, *Expert Systems for Civil Engineers: Technology and Application*, Mary Lou Maher (ed.), ASCE - Books, 345 East 47th St., New York, NY, 1987.

Olsen, B., Pumplin, B., and Williamson, M., The getting of Wisdom: PC Expert System Shells, *Computer Language*, Vol. 4, No. 3., 1986, pp. 117-150.

Ortolano, L., and Perman, C.D., Software for Expert Systems Development, *Journal of Computing in Civil Engineering*, Vol 1, No. 4., October, 1987, pp. 225-240.

Rasmus, D.W., Expert System Shells for the Macintosh, *PC AI*, September/October, 1988, pp. 36-43.

Chapter 15
Selected Homework Assignments[1]

PROBLEM 1. Identify potential applications of expert systems in structural engineering. Use examples if necessary.

PROBLEM 2. Consider a steel column under axial compressive loading. Present your knowledge regarding this column in the form of:

    (i)   rules
    (ii) frames
    (iii) o-a-v triples
    (iv) a-v pairs
    (v) semantic network

PROBLEM 3. Analyze the problem of washer requirements in structural joints using ASTM A325 or A490 bolts (page 6-272, Manual of Steel Construction, Load and Resistance Factor Design). Present these requirements in the form of if-then rules. Use Personal Consultant Plus development shell available in the CE Department.

PROBLEM 4. Suppose that you could develop a computer program in which the format of the language used was in the form of rules such as:

"IF . . . . ,
THEN. . . . ."
     or
"IF . . . . ,
OR. . . . .,
THEN. . . . . . .,
AND . . . . ,
ELSE. . . . ."

Develop a simple set of rules that characterize one of your civil engineering tasks. Be prepared to discuss this task with the class.

PROBLEM 5. Select two papers from the published literature that are closely related to the project which you are considering to undertake. After reading the papers, write a brief summary of each and indicate what aspect of the papers most interested you. Provide me with a copy of each of the papers.

PROBLEM 6. Implement an OPS5 program that will represent the dependency network for simply supported beam design, as shown in Figure 15-1. In the dependency network the variables are nodes and the dependencies are links. You should use production rules to represent the relationships between the variables and the modification of the values of the variables. The WMEs should represent the variables, their values, dependent variables, and status. The values of Sactual and designation are to be stored in working memory.

Your OPS5 program should have the following behavior:

    1. First, the user must specify, by prompting, all leaf nodes except Sactual. The user is then given the values of all other nodes.

    2. The user may then change the value of any node, except Sactual and designation, and be given the values of other nodes consistent with the change.

PROBLEM 7. The object of this assignment is to write a small expert system in OPS5 to give you an idea how to go about implementing a typical diagnosis problem. The task is taken from the domain of car diagnosis. The most difficult part of writing an expert system, that of accumulating and organizing the domain knowledge, is given to you in Figure 15-2. However, since none of the people who prepared the assignment is an expert car diagnostician, please add additional knowledge as you feel necessary. Further, you are given a possible design for the system, but feel free to make any justified modifications.

Problem Description. Figure 15-2 is a sketch of an inference network dealing with battery and starter problems. Also given is factual knowledge about seals that can leak fluid substances (oil and coolant) from one area to another. A possible design of the system is to encode the knowledge embedded in the inference network

---

[1]This chapter was prepared by Mary Lou Maher and Satish Mohan

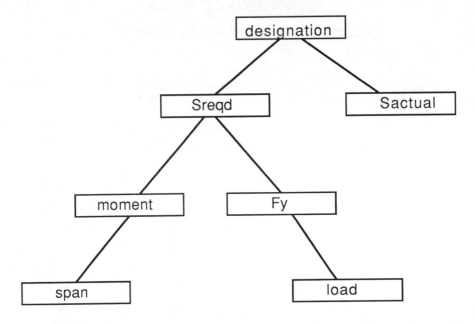

Figure 15-1. Dependency Network for Beam Design (Problem 6)

in the form of OPS5 productions that operate on three classes of working memory elements (WMEs): observables, inferences and conclusions. The knowledge about leakages can be encoded in the form of static OPS5 WMEs. This is convenient because knowledge of seals and leakage is information about the object themselves, while knowledge about other matters consists mostly of heuristic relations.

The first task of the system is to gather evidence. The user is queried about the values of observables. Each observable has a name, a query string, a value, and a status (known or unknown). Make statements can be used to put observables with no assigned values in working memory and one production can then loop through them to get the values from the user.

The second task is to make as many inferences as needed (and possible) to reach the conclusions. Note that not as many productions are needed as there are "boolean boxes" in the inference network. Specifically, the "AND" boxes can be encoded in OPS5 productions, while, most of "NOT" and "OR" boxes can be encoded within those productions without the need to create additional intermediate inferences. The next task of the system is to inform the user of the results of the diagnosis. The system should print to the screen the conclusions that have been reached. Moreover, if a conclusion concerns leakage, some information should be generated to allow some rule(s) to search the "fact base" and identify the possible seals that might be at fault. The name of these seals should be printed to the screen after the leakage conclusion they pertain to.

Finally, if no conclusions can be reached, the system should halt and inform the user that no conclusions can be reached from the given observables.

Note that these tasks should not be done strictly sequentially. The system should start making inferences and printing out recommendations and leakage diagnostics as soon as enough observables have been known. The forward chaining nature of OPS5 allows you to do that without you worrying about an explicit control structure.

Additional Information. The following is a list of the observables that the system has to query about:

1. spark-plug-electrodes-fouled-with-oil *with values* yes *or* no;
2. dipstick-shows-coolant *with values* yes *or* no;
3. jumper-starts-engine *with values* yes *or* no;

SELECTED HOMEWORK ASSIGNMENTS

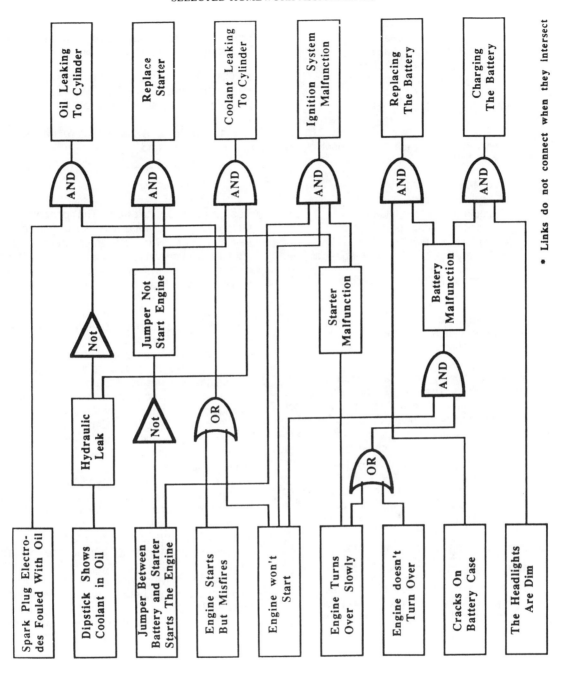

Figure 15-2. Inference Network for Automobile Diagnosis (Problem 7)

4. engine-starts *with values* yes *or* no *or* misfires;
5. engine-turns *with values* yes *or* no *or* slow;
6. cracks-in-battery *with values* yes *or* no; *and*
7. headlights-dim *with values* yes *or* no.

The system could reach one or more of the following conclusions:

1. oil-leak-to-cylinders;
2. replace-starter;
3. coolant-leak-to-cylinders;
4. ignition-malfunction;
5. replace-battery; *and*
6. battery-leads-loose-or-charge-low.

The intermediate inferences that the system might generate are:

hydraulic-lock;
starter-malfunction; and
battery-malfunction.

You will probably notice a peculiar behavior of the system. That is the fact that some answers that the user gives might make other queries immediately irrelevant or redundant. For example, if the engine does start there is no point in asking whether putting a jumper between the battery and the starter makes the car start or whether the engine turns over. You need not do anything about that for this assignment. However, as you will later learn in the course, this problem can be solved by either imposing some control structure on the task or by using a backward chaining inference mechanism.

What to submit You should submit the following:

1. A written description of how your system works. The description should be 1/2 to 1 page of formatted text, at most.
2. A properly commented listing of your program.
3. Two sample runs showing your system performing on two cases of your choice. One of the runs should have watch level 1 and the other watch level 0.

Note. You will be graded on the proper use of OPS5, including taking advantage of the relevant features of the language, on program clarity and readability, on the ease of making additions and modifications to the code and, to a lesser extent, on the efficiency of your program.

PROBLEM 8. The object of this assignment is to redo the problem given in the previous assignment to show you how to:

1. impose a control structure on the problem; and
2. simulate backward chaining in OPS5.

Assignment Description. The observables in the system are divided into two disjoint groups: *key* and *confirming*. The intent here is to first gather key observables only and later ask for the confirming ones if they happen to be needed to prove or disprove some plausible hypothesis.

The first goal of the system is to collect *key observables*. You should add a *type* attribute to your observables that can take the values *key* or *confirming*. For this assignment the key observables will be: *dipstick-shows-coolant-in-oil* with values *yes* or *no*, and *how-the-engine-starts* with values *yes, misfires* or *no*.

The second goal is to select a plausible conclusion from the key observables as follows:

Conclusion 2 *(replace-starter)* and conclusion 3 *(coolant-leak-to-cylinders)* should be selected if the values of *dipstick-shows-coolant-in-oil* is *no* or *yes*, respectively.

Conclusion 1 *(oil-leak-to-cylinders)* should be chosen if the engine misfires, while conclusions 4 *(ignition-malfunction)*, 5 *(replace-battery)* or 6 *(battery-leads-loose-or-charge-low)* should be selected if the engine does not start.

The third goal is to pursue the plausible conclusion to either confirm or disprove it. This will include deriving the values of intermediate inferences and gathering additional confirming observables.

For example, if the hypothesis being pursued is *replace-battery*, then *battery-malfunction* and *cracks-in-battery-case* should be investigated. To get the value of the former, the system needs the values of *how-the-engine-starts* and *how-the-engine-turns*. Since the value of the *how-the-engine-starts* has been asked before, it can be used without querying the user again, but *how-the-engine-turns* has to be queried. If the engine does turn over, then *battery-malfunction* is true, then the user has to be queried for the cracks in the battery case to confirm or disprove the hypothesis.

The system should similarly select each

plausible hypothesis and pursue it as described above. You should be careful not to try to get the values of observables and intermediate inferences that you have already queried or derived while pursuing previous hypotheses.

Finally, when there are no more plausible hypotheses to process, the last goal is to output to the user the results of the session. If some conclusions were confirmed, they should be printed to the screen and the system should halt. If not, the system should express its ignorance before halting.

Additional Notes. You need not worry about inferring the sources of leakages, in this assignment.

You will probably notice that backward chaining at the rule level is tedious in OPS5. You are right. In fact, as you will learn in class, it is typically performed at the task level, where a task usually involves a number of rules. However, because this is a one-week assignment and you don't have the time to encode a large knowledge base, this demonstration shows you how to do it.

What to submit. You should submit the following:

A written description of how your system works. The description should be 1/2 to 1 page of formatted text, at most.
1. A properly commented listing of your program.
2. Two sample runs with watch level 1, showing your system performing on two cases of your choice.

PROBLEM 9. Below is a brief interview made with the manager of a small water supply utility. From this interview you are to develop a simple expert system that operates the reservoir in a manner that is consistent with the comments of the manager. You can leave messages for the manager at (e-mail address).

This assignment is very open ended, of course. Begin your assignment by carefully constructing a reasonable approach to the problem and the limits of how you wish your system to operate. Determine the minimum number of variables that you will need and how they are related. Do not attempt to develop a complex system. This is a weekly home work assignment and should not be viewed as a project. Grading will be based on originality, completeness, and problem approach. I would assume that confidence factors would be very important in this problem.

"The operation of my reservoir generally goes like this: I have a series of rule curves similar to those I have given you. They indicate the range of storages that I desire to maintain. Now, I don't follow them exactly, but I use them as a guide. I am also concerned with the predictions I receive for future streamflows (not that they are so good, in fact, anything over three months I don't believe, two month predictions are not very good, and one month's aren't always correct either).

So anyway, I want to meet the demand that is placed on my system but also keep my eye on the storage levels. If my storage level goes above the top curve, I will release water to get to that level. If it goes below the bottom curve, I will cut back on my release to get up to that level. I also want to look ahead using my streamflow forecasts to ensure that I do not get myself in a situation that will require me to be either above the top curve or below the bottom curve. In fact, if the forecasts indicate that I am going to get in trouble in the future, I may start adjusting my releases immediately.

Oh, did I mention, what I do is I try to establish a monthly release at the beginning of the month and hold to it all month. I don't like to change my policy in the middle of the month. The amount I cut back or overdraft varies for individual months. In the summer I definitely will cut back. In the spring I usually wait and hope for higher inflows. Winter months are real dangerous for floods, so I am really careful. If high flows are predicted I can drop the level of the reservoir quite a lot.

As I said, the flow predictions are not very good. We get them in terms of very high, high, medium, low and very low. The mean values for each of the months are also given on one of the tables I am giving you. That is really about all I can tell you. This is really a subjective job and I don't think anyone could replace me.

For instance, last year I began my storage in January at 45 units and the flows and estimates during the following months were:

	Actual Flow	1 mo. Pred.	2 mo. Pred.	3 mo. Pred.
Jan	69	46	100	158
Feb	72	54	112	174
Mar	65	58	120	172
Apr	75	62	104	142
May	40	50	80	110
Jun	25	35	55	65
Jul	15	20	40	60
Aug	11	20	40	60
Sep	9	20	50	80
Oct	14	20	55	85
Nov	50	30	70	120
Dec	58	40	90	150

The municipal demand during the period was 45 units per month. I managed the entire period with only being a little over the flood curve and a little under the drought curve a couple of times and only having to limit demand occasionally. No sir, no computer will ever replace me."

Can you make a system that can operate for the entire period with no cutbacks and meeting all of the rule curves?

Mean Monthly Flows

January	46
February	54
March	58
April	62
May	52
June	38
July	28
August	14
September	18
October	22
November	34
December	41

PROBLEM 10. Construct a rule-based system that is capable of recommending a herbicide and the appropriate application rate for that herbicide for a given field situation. The user will be expected to supply the following information:

1. soil organic matter content
2. crop to be planted
3. existing weed problem

The problem of recommending a herbicide has been simplified greatly in order to keep it manageable as a classroom example. Information concerning the herbicides and their application guidelines are contained in the following table.

You should provide a menu interface for the user, allowing them to select quit or a new run (initialize the inputs), using the KES HT shell. The crops of concern are corn and soybeans and the weeds are grass and broadleaf. The following truth table should aid development of the necessary rules:

Herbic.	Weed	Crop	Org. Matter		
			2%	2-4%	Over 4%
Sencor	broad-leaf	C/S	Do Not Use	3/4 pt/ac	1pt/ac
Lasso	both	C/S	2qt/ac	1qt/ac	2qt/ac
Bicep	both	corn	1.5qt/ac	2.5qt/ac	3qt/ac

Most harvest losses in corn can be diagnosed by observing two factors: The location of kernels on the ground, and the state of the kernels. Kernels on the ground are generally located in one of three places:

1. in front of the combine where the corp is still standing
2. behind the header, under the combine
3. behind the combine

For this exercise, the kernels will be in one of two possible states:

1. attached to the cob or pieces of cob
2. completely separated from the cob

By observing these two factors, one can determine the part of the combine that is responsible for the grain loss. For our purposes, there are five kinds of loss:

1. header loss
2. snapper roll loss
3. crop loss
4. cylinder loss
5. separation loss

Header loss occurs when kernels are found on the cob underneath the combine. If the kernels are on the ground beneath the combine and are not attached to the cobs, then snapping roll losses are occurring. Crop loss is simply those kernels on the ground in a standing crop (before the combine has passed). A sign of cylinder loss is when kernels on pieces of cob are found behind the combine. Lastly, when free kernels are found behind the combine, separation losses are occurring.

Most of the above losses can be reduced by making the appropriate combine adjustment. Snapper roll losses can be reduced by adjusting the stripper plates on the header. Excessive crop loss is a sign of an overmature crop, which is difficult to adjust for. Cylinder loss can be due to two factors: excessive concave clearance, or a cylinder speed that is too slow.

Header losses can occur because of excessive forward speed of the combine, or misadjustment of the gathering chains. Finally, separation losses can be due to sieves that are opened too wide, excessive fan speed, or overloading of the machine caused by traveling too fast.

In addition to developing a working rule base, use the (dribble-on "*filename*") feature to create a script file of a run of the ES and use the following CLIPS feature before running (after typing (reset)) and after running

  a. (facts)
  b. (agenda)
  c. (matches "*rulename*")

Hand in the dribble (script) file and send me

(engelb) your rule base or path to your rule base.

PROBLEM 11. You are a consultant working for a firm that provides an irrigation scheduling service. You have a number of clients and would like to automate the scheduling procedure. Develop a prototype system using CLIPS given the following information.

The following is known about each field:
name
crop
irrigation status (on off)
soil moisture depletion
amount applied per irrigation

Present a welcome message when the ES is run. Clear the screen using the system command (i.e. (system "clear")). Request the file name containing information for the fields.

The user should supply the corn and soybean ET (evaportranspiration) for the previous day, the rainfall for the previous day, and the chance of rainfall for the current day. (**Note!** corn ET and soybean ET are usually different). An interface to get these data from the user should be constructed. The user should be asked for the field names in which irrigation was completed on the previous day. The irrigation status of these fields should be set to off and the soil moisture status of the field should be decreased by the amount of water applied per irrigation. Each field's soil moisture depletion should also be updated to reflect the current value once the previous day's corn ET, soybean ET, and rainfall amount are known. Rainfall decreases soil moisture depletion and ET increases soil moisture depletion. (Note: A flag on each field might be required to prevent looping once the soil moisture depletion is updated for ET and rainfall.)

Irrigation in a field should be stopped when rainfall causes soil moisture depletion to fall below 0.50 inches. Irrigation should also be stopped when the chance of rain is greater than or equal to 50 percent. Irrigation should be started in a field when the soil moisture depletion is greater than 1.20 inches and the chance of rainfall is less than 50 percent. If the status of the irrigation system is changed the user should be informed.

Once fields have been updated for current conditions, report each field's current conditions to the user before saving.

The field information should be stored in a file and loaded into the fact base when needed. Updated field information should be stored in a file when the program quits. (**Hint:** (save-facts "file")).

PROBLEM 12. Develop an expert system based on a regulatory code for road safety barriers. You will have to deal with the incomplete information provided by the attached Technical Memorandum. Which knowledge determines the type of safety barrier you should select for a particular application? This is the knowledge you want to use to develop your expert system logic (reasoning tree). Which knowledge does not affect the choice of safety barrier, but should be provided to all users of the system? Which is the knowledge known by practitioners in the field, but not by novice users?

Required:
Use the expert system shell INSIGHT2+ to build your expert system and turn in a floppydisk with your files FILENAME.PRL, FILENAME.ICO and FILENAME.STR and a printout of your file FILENAME.PRL.

Incorporate the following features of INSIGHT2+ in your system:

1. In any case, INSIGHT requires you to use: TITLE, THRESHOLD, RULE, IF, THEN and END.
2. Also use AND, value data types "<, or >, or ...", and set queries using IS or ARE.
3. Choose at least two levels of Goals that can be reached.
4. Try out how the ELSE statement works. You'll find out that it is pretty confusing to the user. Please explain why.
5. Use the passive and active provision for explanation: DISPLAY and EXPAND.
6. Try to phrase the rules such that your questions to the user of the system sound like English.
7. Add comment lines in your FILENAME.PRL file to explain what different sets of rules are about.
8. First try your system with the CONFIDENCE set to OFF. Make a choice for the THRESHOLD value. Then turn CONFIDENCE ON and see how that impacts the questions asked to the user. Try the system with various THRESHOLD values. Please turn in the system that you think is best, and briefly explain why it is best.

Background. The following Technical Memorandum provides a set of rules and facts in clear English taken from the Department of Transport Technical Memorandum H9/73. These rules are concerned with the selection of safety barriers for roads in the UK.

The problem is two parts:

1. Establish the need for a barrier.
2. Determine best type of barrier if one is required.

Once the need for a barrier has been established, there are rules to determine which type of barrier is best. The most commonly used barrier is the tensioned corrugated beam barrier. The available deflection clearance will modify the decision of the barrier type. If there is insufficient clearance, another type will need to be selected, or the barrier doubled/section size increased, and/or the post spacing halved.

TECHNICAL MEMORANDUM NO. H9/73 SAFETY FENCES

This memorandum states the criteria for the provision of safety fences in verges and central reserves of roads. It describes the construction and attributes of the different types of fence approved for general use and the circumstances in which other systems may be appropriate. New specification clauses in the 400 Series covering the overall and the materials and workmanship requirements of the principal types of fences are appended.

Criteria for the Provision of Safety Fences

Safety fences are normally recommended only on major roads where speed restrictions permit 80 km/h (50 mph) or above, and for which the circumstances below apply; there will also be a need for safety fences on less important roads where there may be exceptional hazards either affecting the layout or the roadside.

A. Verges

a. On embankments 6.0m or more in height.
b. On embankments where there is a road, railway, river or abnormal hazard at the foot of the slope.
c. On the outside only of curves less than about 850m radius on embankments between 3.0m and 6.0m in height.
d. At obstructions such as bridge piers, or abutments, or posts to large signs.

B. Central Reserves

a. For protection at bridge piers, posts to large signs, etc., which, if displaced would cause danger to other vehicles or persons, but not for small signs which should be constructed as far as possible to give minimum damage to vehicles striking them.
b. For protection at lighting columns.
c. On urban roads without central lighting columns where the central reserve is not more than 1.75m wide and enables adequate clearance to be provided between the face of the safety fence and the edge of the carriageway.
d. On rural roads where the central reserve width is 2.5m or less.
e. On roads where the difference in carriageway levels exceeds 1.0m and the slope across the reserve is steeper than 25% the fence (double sided) to be placed adjacent to the edge of the higher carriageway.
f. On unlighted motorways (including trunk roads to be converted to motorway status) where speed restrictions permit 100km/h (60 mph) or above, and which have central reserves between 2.5m and 6.0m wide, safety fences shall be provided on all future motorways as initial equipment. On the more heavily trafficked existing motorways and those under construction they shall be provided in accordance with the published national program.
g. On unlighted rural all-purpose dual carriageways with speed characteristics and central reserve widths as in f. above, safety fences may be provided on very heavily trafficked roads subject to special justification and specific approval from Headquarters.

Visibility

Where in the road alignment horizontal curvature coincides with convex vertical curvature, safety fences other than the wire rope type may obstruct the stopping sight distance; this effect must be taken into account in the design of new roads.

TYPES OF SAFETY FENCE AND THEIR USE

Tensioned Beam Safety Fences

Tensioned beam safety fences are to be used in preference to untensioned types except:

a. Where the length of fencing is less than 45m between anchorages OR
b. On curves of radius less than 120m

There are two approved types:

i. The tensioned corrugated beam (TCB)

developed by the Transport and Road Research Laboratory and described in Appendix A.

ii. The rectangular hollow section beam (RHS) developed by the British Steel Corporation and described in Appendix B.

The TCB is for general use either in the single-sided or double-sided form on a single row of posts, dependent on the deflection space available or the location in the road cross-section. The RHS is available in two sizes, 203.2mm x 101.6mm and 101.6mm x 101.6mm. These may be used subject to no extra cost being incurred as alternatives to the TCB or open box beam (OBB) in situations where they provide equivalent protection. When the 101.6mm x 101.6mm RHS is used in the central reserve, the surface must be hardened. Contractors should normally be allowed the choice between the TCB and RHS or OBB and RHS but all extra costs of any hardening and levelling of the ground where appropriate must be included. For maintenance and aesthetic reasons, short sections of RHS should not be interposed in otherwise continuous lengths of TCB or vice versa. Because of its slender construction the RHS may be preferred -

i. in areas of high amenity where appearance is important, and
ii. where (especially with the 1001.6.mm x 101.6mm RHS) it is material to reduce the risk of operational delays due to snow drifts on the carriageway.

In such circumstances alternative costs for both TCB and RHS, including any hardening of the ground where appropriate should be obtained so that the amenity or traffic benefit can be judged against the additional costs.

Untensioned Beam Safety Fences

There are two approved types:

i. Open Box Beam (OBB) described in Appendix C
ii. Blocked-out Beam (BOB) described in Appendix D

These fences are intended for use in providing protection at obstructions over short lengths and where space for deflection is limited. Although more expensive, the open box beam safety fence, developed by the TRRL for use on motorways and other high speed roads, is more effective than the blocked-out beam type, both in protection at obstructions and in controlling the path of vehicles after impact. It can be used on vertical or horizontal curves down to 30m radius and can be mounted on bridge piers or concrete parapets by means of hexagonal brackets, so that the face of the beam is 0.45m from the face of the pier or parapet. For a given clearance from the edge of the carriageway this enables wider bridge piers to be provided on the central reserve than is possible with either the double tensioned beam or blocked-out beam types. (See Technical Memorandum H9/71).

A 2-rail open box beam fence which provides added height and strength is available for use at special sites, e.g., a road crossing a dam embankment, where there is an exceptional hazard to the occupants of vehicles which might penetrate or roll over a single height type of safety fence or bridge parapet. (See Appendix C.)

Due to the higher risk of colliding vehicles rolling over blocked-out beam fences, they should be installed only at situations where the consequences of this hazard are unlikely to be severe.

Other Types of Safety Fencing

Christiani and Neilsen

The Christiani and Neilsen safety fence has been redesigned to overcome the deficiencies revealed in the testing of the earlier type. Tests have now shown the new design to be satisfactory for sites where frequent light collisions occur (RRL Report LR 246). It may be used in such circumstances where it is likely to prove economical in maintenance from its ability to sustain frequent light impacts with little damage. Its cost is expected to preclude its use where long lengths of safety fence are required. Both ends must be ramped down to ground level; it should not be connected to a tensioned type of safety fence.

Wire Rope Safety Fence

A wire rope safety fence has also been developed by the RRL in conjunction with British Ropes Ltd (RRL Report LR 98). It is satisfactory provided the correct height of impact is maintained and for this purpose, the ground needs to be hardened and level, in front of and behind it. It is more flexible than the tensioned beam type and normally is unsuitable for protecting obstructions, because it requires a deflection clearance of at least 1.8m. It may be used in the centre of central reserves or on embankments as an alternative to the tensioned beam type, provided the necessary effective clearance is available and the ground is hardened. It cannot be used on curves of radius less than 610 m. The posts must be mounted in concrete footings. When the use of wire rope safety fences is being considered, account should be taken of the added costs of providing beam safety fences at obstructions and of hardening the ground; the latter will be of less importance on all-purpose roads and dual 2 lane motorways where 1.0m hard-strips are provided on the reserve. There are some special conditions where this type of fence may be of advantage, such as where minimal air resistance is needed to prevent snow accumulations on mountain roads. Due to the greater flexibility of these fences they are unsuitable where there are street lighting columns in the central reserve and should not be used where there is a possibility of such future lighting.

Concrete Barriers

A barrier consisting of a solid contoured wall constructed of mass concrete has been developed in the state of New Jersey. Vehicles can run close to the wall without scraping their sides.

Full scale impact tests conducted in the USA have shown that for low angle impacts and speeds up to about 100km/hour the barrier satisfactorily redirects a 4500 lb vehicle with minimal damage to the vehicle and no damage to the barrier.

This barrier should be considered for use in circumstances where its expected extra cost (including hardening of the central reserve) above that of other approved types of safety fence would be justified by the savings in 'land take,' maintenance costs and delay costs, from disruption of traffic during repair. At present the main use will probably be on heavily trafficked urban roads where both road and central reserve width are restricted. There may, however, be other circumstances, e.g., long viaducts, where this type of barrier is the only one possible because of space restriction.

It should be noted that for satisfactory performance the surface up to the barrier should be hardened. If necessary the wall section can be thickened or the barrier constructed in two separate halves to accommodate lighting columns or signal or sign posts. Drainage or cable ducts can also be incorporated in the base section.

Appendix A

Tensioned Corrrugated Beam Safety Fence

General The steel beam is mounted at an overall height of 760mm above the edge of the adjacent carriageway, hard-shoulder or hard-strip on mild steel Z section posts, to which it is attached by shear bolts, and tensioned between anchorages sunk in the ground to which the ends are sloped down. Upon vehicle impact the bolts fracture, allowing the posts to be knocked down without appreciably affecting the original height of the beam. A stiffer fence is obtained by mounting 2 beams, one on each side of the standard post; further stiffening may be obtained by halving the post spacing. The design is the subject of Patent No. 1141012. Aluminum alloy beams may be used provided they are capable of meeting the specified steel beam strength and stiffness criteria.

Clearances Ideally a clearance of 1.2m behind the single beam fence is desirable to allow for deflection on impact. A double beam fence will reduce the space required for deflection to 0.6m; this will be further reduced to 0.46m by halving the post spacing. however, on lighted motorways with columns on the central reserve it has been decided that single sided tensioned beam safety fences sited 1.2m from the edges of the central reserve and with posts at half spacing for 3 bays in advance of each column will suffice as evidence indicates that impact with fences on the central reserve is generally of a glancing nature, and also that fences near the edge of the carriageway may lead to an increase in accidents where vehicles have not entered the central reserve. On all purpose roads with 4.5m central reserves, this layout will provide 0.9m deflection clearance to the lighting column. On motorways constructed with 4.0m central reserves, the deflection clearance to the lighting column would be only 0.6m.

Appendix B

Tensioned Rectangular Hollow Section Beam Safety Fence

General A rectangular hollow section either 203.2mm x 101.6mm or 101.6mm x 101.6mm is mounted at a centre line height of 610mm above the edge of the adjacent carriageway, hard-shoulder or hardstrip on top of the same Z-section mild steel posts as are used for the TCB fence but connected to them by U-straps which are attached to the posts by shear bolts. The U-strap design has been developed by the British Steel Corporation and because the post to rail connection does not involve any holes, slots or fixings to the beam the post spacing can be readily varied to avoid underground obstructions. The two sizes of RHS are for use in different situations depending upon the space available for deflection. Where this is particularly limited the 203.2mm x 101.6mm RHJS may be mounted on the traffic face of the stronger Z-section post as used with the Open Box Beam safety fence or directly on a structure by means of hexagonal brackets. Individual lengths of beam are joined by means of internal mild steel fishplates tapped to receive screws; tensioners are installed at intervals to eliminate slack and minimize deflection, and to accommodate temperature expansion and contraction. End sections are ramped down to ground level and anchored. In a vehicle impact the fence behaves similarly to the TCB. As the beam is normally mounted on top of the posts and presents the same profile to both sides, it performs as both a single and a double-sided fence.

Clearances The following clearances behind the back face of the RHS fence beam are desirable to allow for deflection on impact:

101.6mm x 101.6mm RHS mounted on top of 100mm x 32mm Z-section posts at 3.2m centers	1.2m
203.2mm x 101.6mm at 3.2m centres	0.8m
203.2mm x 101.6mm RHS mounted on traffic side of 110mm x 50mm Z-section posts at 2.4m centers	0.6m

On the central reserve of lighted dual carriageway roads it will frequently be impractical to provide the appropriate clearances to the lighting columns and smaller values may have to be accepted.

Appendix C

Open Box Beam Safety Fence

General The beam is mounted at an overall height of 710mm above the edge of the adjacent carriageway, hard-shoulder or hard-strip on mild steel posts, or on bridge piers or concrete parapets by means of a hexagonal bracket. The design is capable of development as a double beam fence mounted on one line of posts for which additional holes are to be specified in the opposite flange of each post and a second plate welded to the opposite side of the end post.

The fence may be used at wide bridge piers on central reserves where there is insufficient space for deflection of a tensioned beam safety fence. The barrier may be connected by means of a transition piece to the single sided tensioned beam safety fence. This avoids doubling of the tensioned beam at and in advance of and the provision of anchorages on each side of the pier.

A 2 rail fence design, one open box beam mounted at normal height with a second one 410 mm above it, is capable of containing and redirecting a 5000 kg coach impacting at 80 km/h and at angle of 20°. Detail drawings in the series SG 1040.18/B may be obtained from Engineering Intelligence Division. It is intended that this type of fence should be used only in special circumstances.

Clearances To accommodate the deflection it is preferable to allow a space of about 0.6m clear of obstruction behind the beam when mounted on posts. This is frequently impracticable where there are lighting columns and a smaller space is acceptable. (See Drawing SD 1040.16/B - GA5)

Appendix D

Untensioned Block-Out Beam Safety Fence

General The steel beam is mounted at an overall height of 685mm above the edge of the adjacent carriageway, hard-shoulder or hard-strip. It is attached to 150mm by 150mm by 1.8m long wood posts by means of a wood blocking-out piece 350mm high, 150mm wide and 200mm long. Aluminum alloy beams may be used provided they are capable of meeting the specified steel beam strength and stiffness criteria.

Clearances There should be a clearance of at least 300mm between the back of the post and the face of the obstruction being protected.

Acknowledgments. The editors would like to acknowledge the following contributors to this chapter: G. Anandalingam; University of Pennsylvania; Raymond Levitt, Stanford University; Steven Fenves and Mary Lou Maher, Carnegie Mellon University; Tomasz Arciszewski, Wayne State University; Richard Palmer, University of Washington; and Jeff Wright, Purdue University.

Chapter 16
Selected Project Assignments[1]

Several of the expert systems courses emphasize semester project assignments over weekly assignments and examinations. The selection of an application is usually part of the project, rather than assigned. This chapter includes a variety of project assignments, selected on the basis of variety in the instructions or plans for carrying out the project.

PROJECT EXAMPLE 1

You are required to produce an expert system to solve a systems engineering problem. This is a team project, with four students to a team. The project includes producing the software, presenting and demonstrating it to the class, and writing a report not to exceed 20 double-spaced typed pages including diagrams.

Deadlines:

Pre-Proposal	February 11
Proposal	March 1
Presentations	April 14-21
Final Paper	April 28

The Pre-Proposal is a verbal presentation of your ideas for a topic and method of implementing the expert system. The topic has to be approved before you move on to the next stage.

The Proposal (< 5 pages) should include the following:
(i) Overview of the problem (at least one page),
(ii) justification of an expert systems approach,
(iii) overall architecture of the expert system,
(iv) what kind of knowledge you need (at least one page),
(iv) who the expert(s) is (are), why (s)he (they) are suitable for providing knowledge, and assurance from them that they will help you, and
(v) progress to date. Include in your proposal the names and departments of the team members.

Some Suitable Topics
1. Scheduling of manufacturing operations in an industry.
2. Screening job applicants and ranking them.
3. Selecting hazardous waste facilities which should be cleaned up first.
4. Designing the topology of communications networks.
5. Examining simulation output and making judgments on test results.
6. Advising all levels of undergraduates on what courses to take.
7. Examining characteristics of chemicals and providing their hazard potential.
8. Designing processes for changing the product mix in an industry or refinery.
9. Designing specifications for a data network based on user requirements.

PROJECT EXAMPLE 2

Narrative Description of Project. Provide a one-paragraph description of your major and your professional background and interests so that we can put your proposal into proper context.

Provide a brief (2-4 page) description of your proposed KBES. Address the following major issues:

Domain. What specific problem will the expert system handle?

Motivation. Why are you proposing this specific application?

Intended Use. Who will use it and for what purpose? *Note:* address a future production version; don't limit yourself to the prototype

---

[1]This chapter was prepared by Mary Lou Maher and Satish Mohan

version that you will develop this semester.

Sources of domain expertise. Where will the domain expertise come from (e.g., your own previous experience, textbooks, manuals, faculty)?

Expected difficulty of compiling domain knowledge. Are the domain heuristics well known? What difficulties do you foresee in collecting and organizing them?

Functional Specification. Provide an extended functional specification of your proposed KBES. Emphasize the following points:

Control structure. What are the important subproblems? How are they interrelated? What control strategy do you propose?

Context structure. What are the principal objects and their attributes? How are they related? What is given and what is inferred?

Kinds of domain knowledge. Will domain knowledge be largely empirical? Will codes and standards be used? How much procedural computing will be needed?

Analogy, precedents. Use reading material provided and identify some KBES, papers or previous student projects which may serve as useful analogies or precedents for your work.

Summary of Domain Knowledge. Provide an extended summary of the domain knowledge to be incorporated into your proposed KBES. Emphasize the following points:

Sources. Where will you obtain the domain knowledge from? If it is from experts, what arrangements have you made for knowledge acquisition from them? If it is from books, etc., provide a bibliography. If it is from your prior experience, in what form do you have the knowledge recorded?

Verification. Do you have problems you can test your KBES with? Do you have solutions again which to verify your KBES?

Heuristics. What are the principal heuristics? Are they purely empirical associations or are they "rules of thumb" for shortcuts which bypass causal reasoning? Will you need to incorporate codes and/or standards into your system? How much procedural computing is required in addition to the heuristics?

Strategies. What are the strategies for planning and executing the problem solution?

Design Specification. Provide an extended design specification of your proposed KBES. Emphasize the following points:

Control Structure. Produce the design for the control structure of your expert system. Identify all tasks (goals) and subtasks and the manner in which they will be undertaken (forward chaining, backward chaining, etc.). Document your design by drawing an inference network in terms of the tasks and subtasks identified. Discuss your style for implementing the control, whether the majority of control will reside in production memory or in working memory.

Working Memory Design. Produce the design for the working memory organization for your project. Identify the WME classes you will need, and the attributes for each class; indicate what each is used for. Be sure to include both the static data (facts) you will have to consult and the dynamic data (context) you will manipulate. Pay particular attention to grouping closely related data elements together in a WME class, and to the relations (cross-references) among WM elements you will establish. Also remember to group information obtained from the user as separate WME classes from the portions of WM that are inferred by the system.

Sample Implementation. Present the OPS5 implementation of an initial segment of your expert system. Submit the listings of your production and WME declaration files and listings of one or more sample executions. Discuss any change in the sample implementation which differ from that proposed in the earlier assignments.

Prototype Project Report. Present the OPS5 implementation of your prototype expert system. This report should be be a complete description of the prototype system from which others can read and understand what you did. The content to the report can be a collection of the previous assignments with corrections and modifications made to reflect the current prototype. In the appendices attach the listings of your production and WME declaration files and listings of one or more sample executions. If you are getting input data from a file, include that file as well.

The following format is strongly recommended.

1. INTRODUCTION
1.1 Scope
   What is the system intended to do, in layman's terms?
1.2 Motivation
   Why was this system selected and why do you consider it useful?
1.3 Expected Use
   Where, by whom, and how would your system be used?

1.4 Sources of Domain Expertise
Where did you get the domain expertise for the system?

2. PROTOTYPE SYSTEM
2.1 Scope
What part of the scope given 1.1 was implemented, and why?
2.2 Overview
Narrative, non-procedural description of how the system goes about its task.
2.3 Working Memory Organization
Principles used for structuring WM, major classes of WME, etc.
2.4 Control Structure
Principles used for control organization (process knowledge), major goals, tasks, etc.
2.5 Domain Knowledge
Major domain knowledge sections.

3. PROTOTYPE EVALUATION
3.1 Scope
Were the right parts of the task implemented?
3.2 Results
Does the system produce the right output and is the output useful?
3.3 Environment
Was a good formalism chosen?
Was the Andrew environment accessible and easy to use?
Was OPS5 appropriate?
Were the additions to OPS5 (Fill and DTE) useful?
Are there any tools that would have greatly facilitated implementation?
3.4 Extensions to Existing System
Are they worth pursuing?
What part(s) have you pursued?
What part(s) should be pursued further?
3.5 Appropriateness of Expert System Technology
Was the task an appropriate Expert System task?
If not, why not? If yes, why?

4. COURSE EVALUATION
Your comments on the course: contents, organization, resources, staffing, etc. Your suggestions for improvement.
APPENDICES
1. Final System Code Listing
2. Sample Executions

PROJECT EXAMPLE 3

Project Proposal

Guidelines to Select a Topic A good heuristic to determine whether the problem you may choose is interesting enough, and yet not too elaborate, is that it should take an expert more than a few minutes but less than half an hour to solve the problem. If you run out of ideas of what might be a good problem, please see the instructor or TA.

We expect you to develop an expert system with 30 to 50 rules. The grading of the final project will be based on 1. the accuracy of the consultation provided, English, user interface and logical sequence of questions, 2. the quality of the knowledge engineering that you used to structure your system, and 3. how your system performed by external validation.

Required.

1. Describe the problem.

2. Where is the expertise required to solve this problem? Does it really require expertise, or is the main difficulty of the problem an elaborate data search. Are there analytical methods available to solve the problem, or does an expert use a combination of the above?

3. Identify the knowledge source, who will your expert be (a person, a textbook,...)?

4. How does an expert solve the problem, and how long does it take him/her to reach a solution?

5. Who would benefit from your system?

6. Who do you think will the prospective user of your system be?

7. How will you be able to tell whether your system gives "good" answers?

8. Which of the expert system shells will you be using? How did you select the shell ?

Project Prototype

Turn in a printout of your prototype program (say at least about ten rules).
Draw a system flow chart or decision tree

(system logic) and explain how your system follows the expert's reasoning.

How will you further extend the system? What will your final expert system be able to do, and not be able to do?

Now that you established the source of knowledge, how will you validate your system?

Who would agree, and who would disagree with the conclusions of your expert system.

Which features of the expert system shell you choose are most important in your application? (eg. database acquisition, explanation screens with drawings, user's confidence,...)

Next week during the tutorial sessions we will ask you to demonstrate how your prototype system runs.

Final Project

<u>Final Report:</u> Your problem description, about 500 words long, must contain:

1. The precise description of your particular problem domain.
2. Who is the intended user of your system?
3. How will your expert system be used? Is it a decision delegation tool, a checklist, a training aid, or a quality assurance tool,...?
4. What sources of knowledge did you use?
5. How did you validate your expert system internally and externally? Does your system make the same conclusions as your expert(s)?

In addition, provide:

If you used the DECIDING FACTOR: (1.) Turn in a diskette with your program, clearly labeled with the name of the final version and the shell used. (2.) Make a printout of the Detailed Model (option #1), and write your filename on it. (3.) On the model printout, number each rule in which you used a kill-value or conditional logic. Explain in a paragraph how it is supposed to work.

If you used INSIGHT2+: (1.) Turn in a diskette with your program, clearly labeled with the name of the final version and the shell used. Be sure that any Pascal routines or database files are on the diskette! (2.) Provide a Logic Flow Chart of your system.

The grading will be based on the quality of your system, as follows:

1. Depth of Knowledge (50 points): What depth of expert knowledge did you integrate into the reasoning, and how did you structure it, while still asking "objective" questions to the user. Do the questions you ask to the user make sense to him/her? Can he/she follow the logic of the system, and learn to reason in the same way? Can the user easily answer the questions by providing facts, or is the system very sensitive to the individual's interpretations of the questions? We don't want you to put in as many rules as possible; rather make good rules, that capture the essence of the reasoning, and make it easy on the user to provide accurate answers.

2. Clarity of Your Expert System (20 points): Is your system user friendly? Are the questions well-phrased; do you provide appropriate explanations and displays to the user; do you illustrate by means of pictures or tables. Note that if you used INSIGHT (rather than INSIGHT2+) you are missing an opportunity here to custom-tailor your system.

3. Validity of the System (30 points): Is your system internally valid, or does it contain logical flaws. How well do its recommendations correspond to those of the expert or other knowledge sources used to develop the system? How well did your system perform under external validation?

Acknowledgments. The editors would like to acknowledge the following contributors to this chapter: G. Anandalingam, University of Pennsylvania; Raymond Levitt, Stanford University; Steven Fenves and Mary Lou Maher, Carnegie Mellon University.

Chapter 17
Selected Examination Questions[1]

QUESTION 1. This problem deals with the diagnosis of a system composed of an interconnected collection of devices, as shown in Figure 17-1. Each device takes two inputs and produces an output. Values from the output of one device are propagated to the input of another one through wires (see sketch). A device is malfunctioning if its actual output value does not correspond to the expected value of the output. This wrong value gets propagated through the system and results in a final output value that does not match its expected value.

The system is to be tested by giving values to all input wires and observing the value at the final output. If the actual value at the final output wire is not equal to its expected value, then one of two situations exists. Either the last device is faulty if its incoming wires have their expected values; or one of the incoming wires doesn't match its expected value indicating that the fault is downstream. The same procedure is then repeated starting from that wire until the faulty device is identified. Assume that the system can have at most one faulty device.

Write an OPS5 program that implements the above test procedure. You should make sure that you check *only* the devices needed for the diagnosis to proceed. Assume that the input to your program (i.e., the initial contents of WM) consists of a description of the interconnection between devices and wires, the actual and expected values of all wires, and the name of the final output wire. Define your own literalize statements. Your program should identify the device that is faulty or signal that there is no fault in the system.

QUESTION 2. The following problem in spatial recognition requires that an expert system recognize lines composed of individual segments which are geometrically collinear. Segments are defined by two end coordinates x1, x2 (assume x1 < x2) along an axis. The task is to determine whether a set of segments forms a line, i.e., a single continuous non-overlapping sequence (endpoints of segments are shared with adjacent segments) or not.

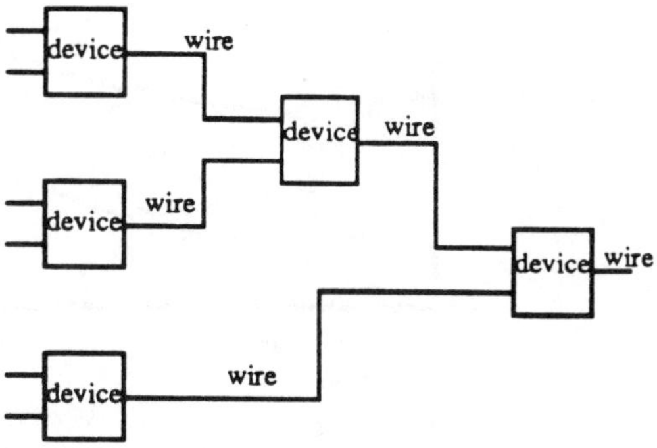

Figure 17-1. Interconnected Devices

Write an OPS5 program which will aggregate a set of individual segments (WME class segment) into a new element line. The endpoints of line should be the beginning (1x) and end (x2) of the aggregated individual segments regardless of how many individual segments are aggregated together. The input to your program consists of an arbitrarily ordered set of segments. If there is an aggregation, the program should output the endpoints of the line and the number of its constituent segments. If there is no such aggregation, the program should recognize the fact (See examples below).

(segment ^x1 5 ^x2 8)
(segment ^x1 2 ^x2 5)
(segment ^x1 8 ^x2 10)
-----> line from x1=2 to x 2=10; 3 segments
(segment ^x1 5 ^x2 6)
(segment ^x1 4 ^x2 9)
-----> input does not form a line

QUESTION 3. Given the inference diagrams shown in Figure 17-2, label the nodes in the order that they will be visited for the four given strategies. All the nodes should be visited once. Which of the four strategies is implemented by OPS5?

---

[1]This chapter was prepared by Mary Lou Maher and Satish Mohan

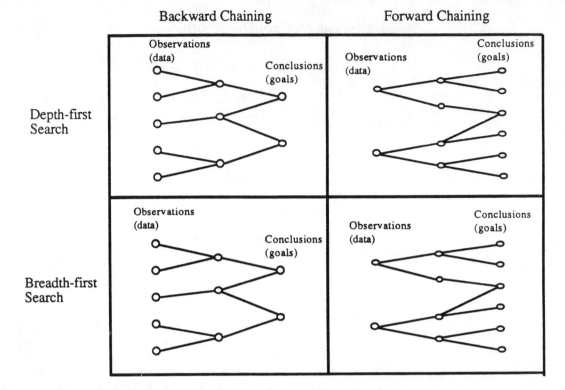

Figure 17-2. Inference Diagrams

QUESTION 4. Write a four-rule commented program in CLIPS or OPS5, that reads in integers, checks the data for consistency and end of file, sums the data to generate a sum, and writes the appropriate answer to the terminal. A sample data file is below:

    1 A 2
    3 4
    5 6 7 8
    eof

QUESTION 5. Describe two advantages and two disadvantages of storing problem solving knowledge in working memory vs. in production memory (OPS5).

QUESTION 6. You are a structural engineer working with a knowledge engineer on the development of an expert system for the selection of steel beam types to be used for different purposes. Present elements of your knowledge related to this area in the form of:

  a. Object - Attribute - Value Triples
  b. Attribute - Value Pairs
  c. Frame

Give at least three example of if-then rules to be used in the expert system under development.

QUESTION 7. One definition of an expert system is "a computer program that is capable of performing tasks at the level of a human expert in one domain." Defend this definition or criticize it (by including your own definition).

QUESTION 8. Give two characteristics of expert systems that distinguish them from conventional programs.

QUESTION 9. What is a production system? Illustrate with a diagram of the components and their interaction. Give two distinguishing characteristics of production systems.

QUESTION 10. Describe forward chaining and backward chaining and illustrate with an inference network.

QUESTION 11. What are the three major components of a VPExpert™ knowledge base and what purpose(s) does each one serve?

QUESTION 12. Explain how backward chaining works in VPExpert™.

QUESTION 13. Describe OPS5 as a production system language.

QUESTION 14. Name two of the three purposes of working memory described in class.

QUESTION 15. What are two of the mechanisms or steps for identifying the dominant rule in the conflict set in OPS5?

QUESTION 16. Why is building a prototype important in developing an expert system?

QUESTION 17. What is the difference between the derivation approach to problem solving and the formation approach?

QUESTION 18. Consider the following fact base

    (forall (x)

    (if (is-sane x)

      (not (inst x AI-course instructor))))

    (inst Alex AI-course-instructor)

    (inst Gerry AI-course-instructor)

    (not (inst Clyde AI-course-instructor))

    (inst Clyde Circus-elephant)

    (forall (x) (if (is-genius x)

        (inst x circus-elephant)))

    (not (exists (y)

      (and (male y)

        (inst y circus-elephant))))

    (male Alex)

    (forall (x) (if (not (male x))

      (female x)))

a. Translate these into English sentences

b. Translate these into the new "language" that contains the following vocabulary introduced in class:

    ----->implies   ~not
    &and   etc.

c. Using this fact base, and backward chaining starting with negations, state if the following formulae are true or false, or cannot be determined.

(a) (not (is - sane Alex)
(b) (is-genius Clyde)
(c) (is-genius Alex)
(d) (forall (x) (if (inst x AI-course-instructor)
    (Not (is-sane x))))
(e) (forall (x) (if (not (is-sane x))
    (inst x AI-course-instructor))))

QUESTION 19. Express the following things in predicate calculus. For each, carefully explain the meaning of each vocabulary item you introduce.

(a) Roses are red; violets are blue.
(b) Every elephant has a trunk.
(c) Every chicken is hatched from an egg.
(d) The coat in the closet belongs to Sally.
(e) Everybody loves somebody sometime.
(f) An apple a day keeps the doctor away.

QUESTION 20. Consider the following fault-finding problem:

"If an engine misfires, and the spark at the plugs is intermittent then the ignition leads are loose or the battery leads are loose. My engine misfires. The spark at my plugs is intermittent. My ignition leads are not loose"

a. Translate these sentences into predicate logic.
b. Put the logical expressions derived above into conjunctive normal form.
c. Show that "My battery leads are loose" follows as a logical consequence of this set of sentences using both forward reasoning and backward reasoning.

Acknowledgments. The editors would like to acknowledge the following contributors to this chapter: Robert Allen, University of Delaware; G. Anandalingam, University of Pennsylvania; Steven Fenves and Mary Lou Maher, Carnegie Mellon University; Tomasz Arciszewski, Wayne State University; Richard Palmer, University of Washington.